The Designed World of Information

Unveiling the Incredible Realm Beyond

Sergei V. Chekanov, Ph.D.

ErmisLearn 2024

The Designed World of Information

Unveiling the Incredible Realm Beyond

By Sergei V. Chekanov, Ph.D.

Editor (general): T. Smaltzar, MD

Editor (English edition): A. Chekanov

Illustrations: A.D. Pascual de Astorza

- Edition 1.0, May 2024
- Edition 1.1, August 2024

ISBN: 979-8-9906428-3-6

ErmisLearn 2024

English translation of the original Russian edition (May 2024): "Неслучайный Мир Информации: Невероятная Реальность за Гранью Нашего Мира" (С. В. Чеканов).

For information about permission to translate or reproduce parts of this book, write to: S.V. Chekanov (author), A.D. Pascual de Astorza (images)

Dedicated to my parents.

Table of Contents

Chance is perhaps the pseudonym of God when he does not want to sign.

Théophile Gautier (1811-1872)

1 Foreword

1.1 From the Author

Are you very sure that the world around you contains no secrets that affect your fate, and that the Universe is not trying to communicate with you?

The world surrounding us is traditionally explained by physical laws and cause-and-effect relationships. There is no need for a God or any immaterial forces to build a machine, make a smartphone, or a particle accelerator. It is believed that if the laws of nature are set, they autonomously work and define the structure and behavior of our Universe. Scientists find such laws and use them to create things that ease our lives.

However, it's not all that simple. Surely you have heard stories about incredible coincidences. They occur without any reason, and their likelihood (often, in subjective opinion) should be very small. For example, you dream of a person with whom you have not communicated for decades, but the next morning you actually meet them on the street. Or you encounter a situation in life that exactly matches the plot of a recently watched movie. Each of us has heard of remarkable predictions and all kinds of omens. After some time, it turns out, some may come true. Your life experience tells you that such events are highly improbable, but you do not know how to explain them.

Such amazing phenomena cannot be explained from the standpoint of materialism. Most often, such cases are rejected by science and considered as statistical coincidences. This is because calculations of the probabilities of such events occurring by random chance are challenging. But what if the Universe is sending us signs or some messages? Could such coincidences in our lives be the product of the unity of mind and matter?

Despite skepticism, such events can be used as a key to gain access to the primary reality that stands behind the scenes of the world stage. To understand such phenomena, the right approach is necessary. In my book I use the laws of mathematics and probability theory to explain specific well-documented events that have occurred with destinies of people. I also show computational experiments that can easily be repeated to illustrate the possible nature of the material world. Or is it just whimsically arranged theater decor, that we perceive as life? It turns out, there are well-documented events involving the fates of people, things, and the surrounding nature that are very rare. They can be perceived as some strange signals, questioning the rationality of the Universe.

Although the word "God" is not present in the subtitle of this book, I hope this discussion may convince the reader that it is God who is behind the arrangement of this world. However, this manuscript is not about religion. In this text, we use the word God as the most concise way to define the common source of life and the meanings carried by information – something you will learn about by reading this book.

The word "God" is the most familiar term to the Western ear for denoting the supreme being who created the Universe and life but does not have a material form. This concept resonates with the word "Dao" in traditional Chinese philosophy. According to Daoist teachings, it means "spiritual path", "the profound and eternal foundation of all things" or "superconscious, formless origin of the creation". Dao is the inner essence of the material world and its invisible beginning. Richard Wilhelm (1873 — 1930), a German orientalist and sinologist, brilliantly translated Dao as "meaning". In his world-renowned manuscript "Tao Te Ching", the famous ancient Chinese philosopher of the 6-5 centuries BC Laozi wrote:

"There is something formless yet complete, that existed before heaven and earth .. It depends on nothing and is constant. All-pervasive and unshakable. The thought arises that it is the mother of all things that exist under heaven. I do not know its name, but I call it 'Meaning.' If I had to name it, I would call it 'The Great.'" (Ch. XXV.).

The equivalence of meaning and God is also seen in the Christian worldview. One of the first lines of the Gospel of John in the New Testament of the Bible reads: *"In the beginning was the Word, and the Word was with God, and the Word was God"*. As we know, "word" is the basic independent unit of language, denoting a concept, definition of objects, their properties, phenomena and relationships with each other. It is a unit of information. In modern language, information is data about the state of something. Data can be presented in various forms and have some meaning (or idea) for the recipient.

The ideas of this book largely complement the concept of synchronicity or synchronism — a term introduced by the famous Swiss psychiatrist and founder of analytical psychology, Carl Gustav Jung (1875 — 1961). Synchronism is a creative principle operating in nature that organizes events in a "non-physical" (causeless) way based on their meaning. In my work, I show new facts that support this phenomenon and supplement them with the results of calculations using probabilistic methods. I will demonstrate events in the history of people that can be interpreted as "waves of ripples" in conceptual terms, affecting the destinies of people, associated circumstances, dates and objects related to their lives. Might these observations uncover the fundamental nature of our existence?

I wrote this book for skeptics. There are many studies appealing to the spiritual beginning, faith in God, religion, to convince people of the existence of a primary reality affecting our lives. Here I apply a new concept based on statistics, which is closest to the scientific approach to understanding such phenomena. I provide examples from life and supplement them with computational methods confirming my position that such events are indeed very rare and deserve attention.

And one last note. This book contains some elements of numerology and related signs. For example, in many places in the text, you will find the number 6. This is the most frequently occurring number. The next most popular is the number — 9. Funny enough, by some incredible coincidence, they make up the year, month, and date of my birthday.

1.2 Features of this book

This book is intended for a wide range of readers. Although it contains some strict calculations towards the end, we appeal to them only when necessary. And only for those who are doubtful and want to redo the calculations. This is what makes this work unique.

Another unusual feature of this monograph lies in the combination of concepts of science, philosophy, mysticism, and spirituality. The last three help clarify the situation in cases where science and common sense become powerless.

I have attempted to find an explanation for the amazing coincidences that are quite well documented in the past. All these historical cases have a small probability to occur. This is confirmed by strict mathematical calculations, whose computer code can be viewed in the

Appendix. To verify these calculations, one simply needs to run the software code on a computer and check the answer.

In the grand scheme of things, science says that all these events happened because there is always a small chance that something will happen. Either that or there is no explanation at all. Other possibilities are impossible, as the materialist paradigm does not allow for the existence of design, trying to explain this world through the causes and effects of unconscious forces.

I must also note that any coincidences in the proposed explanation of unusual phenomena are also pure coincidences. With high probability, I can say that many thoughts expressed in the pages of this book could have occurred to many others, considering that the number of people on the Earth is approaching 8 billion. Just the law of large numbers at work. Or is there something else? After reading this book, you will be able to answer these questions. If you find your thoughts on the pages of this book, then this is not just the theory of probability or blind chance leading to a coincidence with what is written. Such thoughts must have occurred to very many, if this book correctly describes our reality and makes the right assumptions, one of which is that the numerous coincidences in everyday life are not entirely random. They are a sign of something common that unites all of us. Therefore, I apologize to those who have already thought in the same direction as me but did not dare to put similar ideas into words on paper.

1.3 Acknowledgements

I dedicate this work to my mother, Nina, whose unexpected death motivated me to gather information for this book. Like me, my mom was not religious. She went to church just in case, usually to light a candle for health and peace. I don't remember her ever speaking about the existence of God. However, the events that occurred during and after her death were so statistically improbable that I came to fully realize that this world cannot be perceived as a reality based on inanimate matter. I think the events after her death were the most convincing demonstration of a different informational reality. I will talk about these events on the pages of this book.

Also, I dedicate this book to my father, Vladimir. He did not go often to church, just like our whole family. Religion in Soviet society was presented as an outdated and archaic phenomenon, not allowing to "move towards communism and a bright future". It was not a part of our daily life. But it was my father who reminded me that this world is full of inexplicable phenomena beyond traditional explanations. One just needs to be open to them. He always liked to say, "Look ... there must be something here", after finding some unusual news.

Having built a career as a physicist working in the field of elementary particle physics and having written hundreds of scientific papers, I attempted to apply computational methods to some known events. This book can be seen as a kind of exploratory journey into the world of unusual hypotheses. I am eternally grateful to the countless number of my colleagues I have met over my 30-year career in physics, with whom I have had many fascinating discussions about the incredible strangeness of this world. They all, without a doubt, have had a tremendous impact on this work. I am also thankful to those who disagree with my assumptions. It is through debates that truth, which we all seek, is born.

I am particularly grateful to Larry Sanger, Tim Chambers, Cosmas Zachos, and the entire editorial team for fact-checking. As I have mentioned before, my goal was to describe scientific facts as accurately as possible. However, their interpretations are left to myself.

They are undoubtedly beyond scientific inquiry and represent my personal opinion. They constitute eternal questions, the answers to which depend on worldview.

Many of my friends, when they found out that I'm writing a book about the phenomenon of synchronicity, sent me instances from their lives where random coincidences seemed very unlikely. Unfortunately, I could not include all these life situations. The plan for this book was more ambitious than just listing cases of synchronicity.

I am grateful to my sister Natalia, who helped with the collection of information and prompted me to make some calculations for events that were difficult to explain from the perspective of a materialistic understanding of the world, where everything works like mechanical clocks, and random chance rules over events whose origins we do not understand.

Above all I would like to thank my wife Tatiana for her love and constant support. She helped to give my manuscript a more literary form, which made it easier for a wide range of readers to understand the material.

I am also thankful to my sons, Alexey and Roman, for their critical remarks from the perspective of materialistic skeptics. Discussing some chapters with them reminded me of my youth and my arguments with my father about the nature of the Universe and the meaning of existence. And, as is often the case, such heated discussions help in finding suitable terms to explain complex concepts and phenomena.

Sergei V. Chekanov (Ph.D.)

(Chicago, February 2024)

2 This world is such because we are in it

The Universe we observe, from elementary particles (with a size of 10^{-18} meters) to clusters of galaxies (sized up to 10^{23} meters), is incredibly complex. But at the same time, each of its separate parts is precisely adjusted to each other. The dominant scientific concept describing this world is called the "fine-tuning" of the Universe. According to this concept, the foundation of the Universe is based not on arbitrary laws, but on strictly defined values of fundamental constants that are part of physical laws. Our world would cease to exist if one physical constant changed by a fraction of a percent.

Let's provide an example. If we change the mass of a neutron — a particle that is a fundamental component of atomic nuclei — by just one thousandth of a percent, it would lead to the instability of the hydrogen atom. This, in turn, would lead to the absence of hydrogen in the Universe — the main building material for stars. As a result, stars would no longer form.

There are about twenty different fundamental parameters defining the structure of our Universe, and this number is constantly increasing with each new discovery. These constant values are interconnected through laws and mathematical relationships. Changing one constant by an infinitesimally small amount leads to changes in other constants. Such an incredibly precise adjustment of physical and chemical laws is necessary for the existence of stars, planets, and galaxies. Moreover, the laws of nature themselves appear as if they were specially selected so that our Universe and life itself could exist.

This precise adjustment is also evident at the planetary level. Minor changes in the size of the Sun, the Moon, or even Jupiter would lead to the impossibility of life on Earth. As a result, humans would never become the witnesses of this Universe.

Currently, the most popular explanation for such an impressive adjustment of physical constants describing our world is based on the "anthropic principle". Its essence lies in the idea that the existence of the Universe is impossible without an observer (a human). It is the observer who comprehends the cosmos with all the necessary qualities for the development of not just biological, but intelligent forms of life. In the infinite diversity of other universes that existed previously or exist (possibly) now, it is only in this Universe that humans appeared, and it is they who observe it. Thus, on one hand, the anthropic principle provides an explanation for the structure of our Universe, the fine-tuning of physical constants, and cosmological parameters necessary for the emergence and existence of intelligent life, while on the other hand, it suggests the possibility of the existence of other universes with different laws (and observers).

The term "anthropic principle" was first proposed in 1973 by English physicist Brandon Carter, however, the idea itself had been expressed multiple times before. Probably, it was first clearly formulated by Soviet astrophysicist Abraham L. Zelmanov (1913 – 1987) in 1955. Unfortunately, many scientists do not recognize this idea as absolutely scientific, as it explains "the unknown through another unknown by the logic of a vicious circle", as the Polish philosopher and futurologist Stanislaw Lem (1921 – 2006) figuratively expressed in his writings.

Another version of the "fine-tuning" explanation is the theory of cosmological natural selection, proposed by the American theoretical physicist Lee Smolin. According to this model, when a new universe arises, it inherits the laws of physics and fundamental constants from its "parent" universe, but with slight random deviations from the original values. Those universes, whose laws of physics do not allow for the formation of stable systems, do not leave "offsprings". Thus, a kind of cosmological "natural selection" of universes takes place.

Unfortunately, even these hypotheses are not scientific, as they cannot be tested. Moreover, this is an attempt to explain one unknown, namely "fine-tuning" of a single universe, through very complex manipulation of other unknown (and unobservable) mechanisms of infinite numbers of (unknown and unobservable) universes.

Here is another unscientific hypothesis explaining why the laws of nature are arranged in such a way that biological life and human existence are possible: Maybe the design of laws and their fine tuning resulted from conscious activity? We plan to discuss this latter idea in this book. But before doing this, it is not unreasonable to ask: Is it bad that all these hypotheses are unscientific?

Science is a specific field of activity aimed at developing and systematizing objective knowledge about reality based on cause-and-effect relationships. The hypothesis that the world was created by intelligence is not scientific. But many known hypotheses involving scientific terms and acting on an infinite number of universes are also not scientific, as they can never be tested. Alongside this, they are too complex compared to the hypothesis of the creation of the Universe and the intelligent foundation on which this world is built upon.

Our Universe is not understood through science alone. Science is just one of the tools for answering fundamental questions, such as "how is it structured?" and "what processes are required to explain this?" without involving any intelligent influence from outside. Not all significant principles can be reduced to scientific-rational ones. There are many other ways to understand the world. Is it true that you had a great-grandfather who loved your great-grandmother? You can reach the truth, but science has nothing to do with it here. Science, by its nature, operates within certain boundaries. It refrains from making judgments on aesthetics and morality, avoids speculation on the ultimate causes or meanings and limits its conclusions to the workings of natural phenomena. Historical documents, logical reasoning, and philosophical inquiry provide valuable resources for addressing questions beyond the scope of scientific investigation. They can also lead to truth. And then there are intuition and feelings, which often come from "nowhere", but can still lead to correct answers.

But let's return to "fine tuning". I invite you to step back and view ourselves from an external perspective. Imagine a spaceship, flying through the depths of space. It is incredibly complex and fully automated. It was created with one purpose — to fly in empty space and support human life inside. The ship consists of a million parts, precisely adjusted for each function. Inside the ship, there is an infant. The ship is built in such a way that it replaces his mother. It raises him, feeds him, and teaches him basic logic using games. The child does not realize where he is and what he is flying for. Eventually, he understands that he is flying in some sort of craft and notices what it looks like from the outside. He starts asking questions. Where am I? What is this? To understand who built the spaceship, he begins to experiment and assembles a similar toy ship from Lego bricks. The child is happy, but here's the problem — the ship made of blocks does not have a single function of the real ship. Trying to recreate the ship's processes, he puts his Lego ship in a bag and starts experimenting with the contents, creating pressure, friction, and even casting spells. Naturally, all his attempts turn out to be futile. The boy quickly comes to the conclusion that the ship was created by a more advanced mind.

His second realization is related to the understanding that the creation of such an amazingly complex environment surrounding him is connected with achieving some mysterious goal, a part of which he is also. Simply, this conclusion follows from his observations — every complex part of the ship is created for some purpose. Nothing complex can emerge out of "thin air". Realizing the uniqueness of his ship, the child begins to guess that he has a purpose too, or mission.

In most cases, his reasoning about his purpose will be correct. Someone, for some reason, created a complex spaceship to support his life. Of course, there may be other explanations. For example, he may have ended up on the ship due to some coincidence. But this will clearly be too contrived, and we are looking for a simple and most natural explanation.

27

Science will not embrace the discovery that this child has made. Intelligent design of complex systems is excluded from scientific explanations. If the child was told that he must use the scientific approach and nothing else, then he would beat and knead the bag with Lego bricks until old age, trying to find a random process leading to functional complexity, but he would never prove the method by which the ship arose through random processes.

From the perspective of materialists, the example I provided is not convincing, as we already know that spaceships are made by beings with intelligence. But is this really important? How else can we express opinions on any matter without relying on the concepts we possess? Even if it were not the described ship, but something more complex, unlike any object known to us before, the conclusion would still be the same. The chance that a complex environment supporting life assembled itself by chance, and the infant simply happened to be in this environment and was able to comprehend this coincidence, is negligible.

Let's now come up with a story in support of "fine tuning" from purely materialistic viewpoints. Can you do this? Even if you manage, it's unlikely that anyone will understand you. After all, you will have to operate with complex and contrived categories for which there is no slightest reason to be true. Most importantly, you need to invent a natural process that creates the incredible complexity of a certain structure performing some functions.

Materialists believe in the creative miracles of the mechanical interactions of molecules leading to life. They require billions of years and vast spaces to convince themselves and others that such a creative power of inanimate matter is possible. This is the only redemptive explanation for such reasoning. Later in this book, we will show that even events that have occurred over several hundred years sometimes do not lend themselves to a mechanistic-naturalistic understanding of the world. Then what can be said about billions of years on gigantic spatial scales? Anything can happen. And here all our expectations that the natural phenomena of nature can rationally explain the incredible complexity of life come to an end.

Our world is arranged in such a way that the truth lies in the simplest explanation. As the theologian and philosopher Anselm of Canterbury (1033 – 1109) said in the 11th century:

"Everything complex needs for its existence that from which it is composed, and to these things, it owes its existence; for whatever it may be, it exists through them, and they do not exist through it; and therefore, the complex can never be the highest."

In other words, a complex explanation depends too heavily on many unpredictable factors. This reduces the likelihood of a correct interpretation of historic events. Complex random (and non-random) processes occurring over a vast period of time can lead to any outcome. It's far from obvious that such natural processes are the reason for our existence. But more on that later.

But what is the reason for the fine tuning of physical constants, laws and the very existence of these laws? The most probable and logical explanation is an initial plan, as in our example with the child and the ship. For every plan, blueprints and calculations are necessary. Information is needed. And also required is conscious decision-making and the ability and the will to execute such projects.

3 Information, meaning and consciousness

If you ask what "information" is from someone dealing with this subject, the answer would be something like this: Information gives details about the state of something, embedded with some kind of meaning (idea) for the recipient. Or information is knowledge that is necessary for orientation and interaction with the surrounding environment.

The problem with all these definitions is that they operate with difficult concepts such as "details" or "knowledge". This does not make the task of defining information any easier. Some even use words like "stimuli" or "meanings", which have value in a certain context for the recipient. In any case, this definition is quite abstract and can be challenging. In most cases, people try to imagine bookshelves, computer hard drives, or something else that contains data records.

In reality, the concept of information is an abstract notion. Information is a fundamental product of consciousness that cannot be touched or seen (Colburn 2000). It is different from thought or knowledge. Thoughts can be organized and fleeting. Knowledge implies awareness or the understanding of a subject based on received information and thoughts. Knowledge is a particular kind of information.

Information can be stored as organized data that can contain descriptions of objects, knowledge, feelings and instructions for performing actions. These are data that make sense to someone. In an abstract form, information does not need a material medium. However, it's much easier for people to understand the term "data", i.e. when information is entered and stored in some material storage. Such data carriers can change and become more complex as society progresses. Initially, these were arrangements of stones on sand, birch bark and papyrus. Now we use paper, modulated electromagnetic waves or computer hard drives. Logically, before information was transferred to material objects in the form of messages, it must have existed somewhere.

Perhaps, it is worth examining what consciousness and intelligence are, as both concepts are central to our discussion. Typically, consciousness refers to the notion of knowledge and perception of oneself as something independent from the external world. Consciousness is aware of its existence and experiences it sensually. It has a perception and understanding of the surrounding world and itself. This is only possible after being separated from the world. Finally, consciousness has free will and can make decisions.

The word "consciousness" will be used interchangeably with the concept of a "soul". In some religions, the soul is that part of us that transitions from one life to another. On the other hand, consciousness refers to the current embodiment of the soul on Earth. For some spiritual thinkers, all information accumulated by consciousness is erased after death, while the soul continues its journey. I cannot say whether the destruction of information actually occurs or not. In my opinion, the obliteration of an incredible number of records with event happenings, accumulated over a person's lifetime, seems quite illogical. If such a thing happens, then what remains for the soul? This book is not religious, so we will not make a distinction between consciousness, the soul and even the "spirit" (the immaterial part that connects us with the celestial realm or the God).

Unlike consciousness, intelligence is the ability to process information. It can also solve problems and perform calculations to manage the environment surrounding a person. It is not related to the sensory aspect of perceiving the world. In this book, we will also use the word "mind" — a combination of consciousness and intelligence.

In this world, material carriers are needed for the existence and exchange of information. It is said that information is encoded onto these carriers. But for information and encoding, intelligence under the guidance of consciousness, which makes the decision to act and launches the required computational algorithms, is required. Intelligence does the main work of processing and encoding the acquired knowledge from messages. At the same time, a second mind is necessary to read and perceive the information. This reading is possible if this second mind has already agreed with the information source how the records should be recognized. Or it itself figured out how to read and to understand the information due to the similarity of its mind's construction or even because of a common source of origin of these minds. Figure.3 illustrates the concept of information and exchange between individual parts, which can represent people.

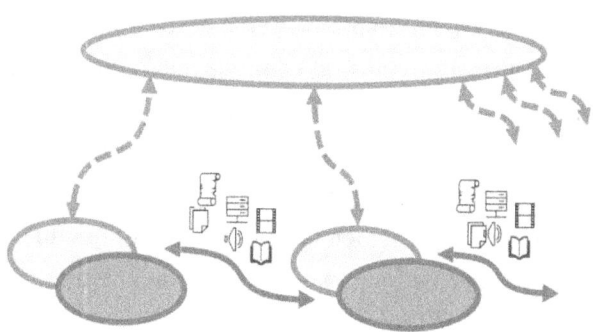

Figure 3. *Dark blue arrows indicate the exchange of information in the form of data. Dark ovals represent intelligence. Light ovals signify consciousness. Dashed arrows illustrate the connection required for the perception of data as some form of information.*

Information in this world always leaves a trace in the form of complex regularities, objects, structures or processes that are easily distinguishable from the typical objects and phenomena of inanimate nature. Intelligence is involved in creating such exchanges through material objects and processes. But most importantly, information is inseparably linked to consciousness, which, with the help of intelligence, can be processed and understood. This is because it is consciousness that abstracts itself from matter.

Suppose you saw these two drawings in the sand:

.. — . .. — . — .. / . —— . — — .. — . (1)

and

.. .. — . — .. — . / . — . — — — . — .. (2)

Which of them represents some information? Until you understand what is before you, these drawings are just certain regularities that could easily have arisen due to natural phenomena, such as wind blows and erosion. These patterns are complex and could potentially represent some recorded information on a physical medium (like sand), until you prove that what you see is some information.

As a matter of fact, the previous drawing (1) carries information. Specifically, it is an instruction to do something. This information is recorded using Morse code — an encoding method in which letters of an alphabet and other symbols are represented as sequences of dots and dashes. This statement means "find water".

However, the pattern (2) means nothing, although it is amazingly similar to (1). How can you tell where the information is and where there is just a random set of lines and dots? Technically, there is no way to do this, until you analyze how these drawings were created, and who was responsible for them. There are several ways when regularity becomes information for us:

- You can read recorded data because you were in contact with the author and learned how the information is created.
- You can decipher data, having some understanding of the authors, their intentions, how they thought, what they might say. In other words, you had some sort of contact with the creators before you saw this regularity.

- You can notice that this regularity has some sort of effect on various actions of mechanical mechanisms, complex devices, or even some (intelligent) beings.

In all these cases, to recognize information, you need to have direct or indirect representations of those who created such regularities. You need some kind of intellectual "bridge" that connects you with the author. It has been observed that if two consciousnesses share a common source and operate with similar ideas and meanings, this is most favorable for the first and second methods. That's why <u>Figure.3</u> illustrates that the consciousnesses of the separate parts (small ovals) are connected or have direct access to the main source of consciousness.

So, information is a product of consciousness. But what about devices and computers that perform certain processes, like collection of data, calculations, and creation of a description in binary code on a hard disk? Do they create information? The answer is "no". They may create data and "modulate" properties of material so it can hold records. But they must be understood. As soon as this understanding occurs, the data becomes information. The loop is completed: Consciousness creates devices to aid in data collection and devises a way to record these data, with subsequent reading and realization. All these devices for collecting and processing information are an extension of our intellect. They are just tools.

We haven't discussed intelligence — the ability to apply information to solve problems, perform calculations and manage the surrounding environment. Can one say that intelligence alone is sufficient for assimilating information? From the example with the tool and the computer — the answer is obvious: Intelligence is capable of remembering, structuring information and using it for computations, which is what computers do. But only consciousness can understand it, make decisions and command the solving of a problem. It can distinguish and recognize itself in the external world as something completely different from matter, which is what consciousness is. We will talk about this later.

What sets computers in "motion" with all kinds of artificial intelligence? If consciousness is necessary to initiate computational processes of intelligence and decision-making, then for launching algorithms in computers, no consciousness is needed. In a microprocessor, some algorithms launch others. Everything happens through calculations, much like in a sophisticated calculator. Perhaps these computations are enough to call the consciousness of biological organisms simply advanced calculations or a byproduct of complex intelligence? It is worth noting here that commands to launch algorithms in a computer are executed by software code. This is a "snapshot" of the human mind, which includes consciousness itself. The computer simply reproduces such code like a record player playing a vinyl record. Software code is the "frozen" in time consciousness of a human being. Perhaps, in turn, human's consciousness is a tiny piece of ice, a frozen in time and space drop of everlasting "water" representing the consciousness of the original spirit of the Universe or God.

But let's return to the information. It is often said that there are computers which, using certain algorithms, can determine whether a sequence of signs constitutes information. One such algorithm is to calculate entropy, which will be discussed in the next section.

3.1 Entropy

Entropy is a value that is used to characterize data or some system. It is the measure of randomness or disorder. Entropy is numerically equal to the number of chaotic jumps in the data or the unpredictability of the occurrence of any character. In this sense, entropy can reflect the amount of information per symbol in a message. However, any random sequence of data without any meaning can also have high entropy.

A set of numbers and words mixed randomly has the highest entropy. The entropy of a repeating word or letter equals 0. If we calculate the entropy of the phrase 'hello world' and then shuffle all the letters, the entropy will be exactly the same as for the original phrase. But as soon as we add an additional letter, the entropy increases. If we take a large number of random letters, the entropy will be the highest. Repetition of identical letters reduces entropy.

In <u>20.1 Appendix</u> we present the calculation of the so-called "Shannon" entropy for three sentences of characters. Here is the result of these calculations:

- "I LIKE SNOW" - Entropy= 3.095
- "I LEKI SNOW" - Entropy= 3.095
- "I LEKI SNOWMAN" - Entropy= 3.378

The first sentence makes sense ("I like snow"). The second sentence does not make sense, but it is made up of the same letters as the first. However, the entropy did not change. In the third case, we added a few letters, which increased the entropy. But it did not add any meaning.

The biggest problem with entropy is that it cannot say whether the text is a message or just a random set of letters. Usually, a message should not have very high entropy, as high entropy corresponds to a random set of diverse symbols or characters. At the same time, low entropy may correspond to a "text" that has no content but simply contains a periodic repetition of certain letters.

To decide whether a set of letters is some kind of message, intelligence and consciousness are required, which can correctly perceive the message. Various algorithms and neural networks can also be applied, which have already been tuned to known messages and therefore can catch an ordered sequence of letters. Nevertheless, they are just tools, created so that a person can understand the meaning of the message. Only consciousness, with the computational abilities of intelligence, can recognize and understand information. It is consciousness that gives meaning to deciphered data and initiates the computational algorithms of the mind. Consciousness is also necessary for creation of information. The intelligence of a machine or a human alone is not enough.

Our Universe can be viewed as a closed system where entropy always increases. This means that all matter inevitably moves towards degradation, disorder and destruction. Consequently, the information contained within our Universe constantly decreases. The very beginning of the Universe was characterized by the maximum amount of information. The information content was higher at the beginning of the Universe when the entropy was lower.

Biological organisms are like small "bubbles" where there is no natural law of increasing entropy during their short existence. Living creatures are vast repositories of well-structured information compared to the surrounding environment. But they "collapse" over time and die, under the influence of the surrounding world, which moves towards chaos.

3.2 Information is easily recognizable

Despite the complexity of the identification of information through computational methods and algorithms, recognizing the signs of information embedded in material objects is not difficult for a human.

Imagine you arrive on a planet devoid of life in a spaceship. After descending the ship's ramp and walking a few kilometers, you discover an object with an incredible number of gears, notches and a vast number of interconnected tubes. What would you think about it? Even without any analogies, the last thought to come to your mind would be that this object arose due to some natural processes on this planet. This artifact must be an informational product of some mind that, using abstract thinking, designed this mechanism and transformed matter into its parts. There simply are no other options. In such situations, there is no place for a materialistic explanation for the appearance of such a complex artifact.

We judge information by the traces it leaves in the material world. Understanding these traces allows us to infer the intentions of the mind that produced this information. Machines built on the basis of informational computations cannot emerge by themselves, no matter how complex the processes they undergo as a result of random collisions of atoms and molecules. No matter how much you shake stones or heat any combination of gases, you will never get a functioning complex object.

Information cannot arise from random processes. Random processes can lead to systems with very high entropy (for example, chaotic mixing of letters) or with low (with some periodicity). Physical processes can indeed create small-scale periodic systems with low entropy and even create systems that may seem complex. However, this would be an apparent complexity that has nothing to do with the presence of information in such structures.

Figure 3.2. *What is the difference between a broken (non-functional) clock mechanism and a group of snowflakes?*

For example, what is the difference between an old non-functional clock mechanism and a bunch of snowflakes shown in <u>Figure 3.2</u>? Indeed, they have such a similar periodicity of structure! Only intelligence and consciousness can resolve this question. Perhaps a mind can conclude that the teeth of gears can interact with each other. This means that some function is possible. Consequently, it is a mechanism.

In this example, we see potentially functional complexity, in which each part performs some action according to a certain process. This is an indication of the presence of information in the design of the entire system. Even if one gear emerged by chance, all other gears must have a specific structure so that their teeth fit together. If the group of gears came into being by chance and chaos, then this randomness must be accompanied by incredible coincidences. Blind chance should act coherently on all its parts. Only then the entire system becomes a working mechanism. However, a working mechanism does not yet mean that there is a certain purpose.

The final conclusion about the presence of information in the design can only be made by consciousness, using the intelligence necessary for analyzing observations. In the case of snowflakes, such functional complexity cannot be found. Therefore, the pattern of snowflakes is just a regularity of structure that occurred according to some natural law. As we mentioned earlier, functional complexity is only one type of the presence of information.

Objects and patterns can also contain messages. For example, a certain regularity on the surface of the sand can be a message, as we discussed earlier. Intelligence and consciousness can determine whether such a structure is a message. But, as we said before, only consciousness makes decisions to run algorithms of intelligence and assign meanings while intelligence helps to carry out the necessary calculations.

3.3 About probabilities

In this book, we will be using mathematics at the most elementary level. For our calculations, we will limit ourselves to dividing and multiplying numbers.

As many already know, probability is the ratio of the number of favorable events to the total number of possible events. Favorable events are the suitable outcome options for an event. If we toss a coin and want it to land on "heads", then the coin landing on "heads" is what we call a favorable event. To calculate probability, we also need to know the total number of potential or possible events. In the case of a coin, there are 2 possible outcomes: "heads" and "tails".

The definition of probability often arises from the notion of equal likelihood of all possible events, which is established based on general considerations of symmetry of the phenomena being studied. For instance, if we have "N" possible events and we have no information indicating that some outcomes are more likely than others, then the probability of any one event out of all possible events would be equal to "1/N".

Let's calculate the probability of getting a "head" (or a "tail") when tossing a coin. In this case, there are two outcomes: It lands on "heads" or "tails". These are equally likely events, as we use geometric considerations — the symmetry and geometry of the coin. There is one favorable event and two possible outcomes. Therefore, the resulting probability is 1/(1+1) = 0.5.

Let's give an example. Imagine that we flipped a coin 5 times and we see "heads" come up 5 times in a row. The probability of such an occurrence is 0.5×0.5×0.5×0.5×0.5 = 0.03125. Or we can write the product of probabilities as $(0.5)^5$. If we ask what the probability is of getting "heads" 5 times in a row for 10 people, then we just need to multiply 0.03125 by 10. We get a probability of 0.3125 (or 31.25%). For 100 people, on average, roughly 3 people will see "heads" from 5 consecutive throws.

However, the event of getting "heads" 10 times becomes very unlikely for 100 participants in the tossing — the probability of such an event is about 0.000976. To see at least one positive outcome requires 1000 participants. And this event will not even be 100% guaranteed for such a large group of participants.

To understand the arguments described in this book, we need to review a bit of elementary probability in general terms. We need to remember how to calculate the probability that two (or more) events occur. In this book, we will deal specifically with such independent events that occur simultaneously, as they always lead to small probabilities.

Turns out, it's all quite simple. If you know the probability of one event and the probability of a second event, you simply multiply these probabilities. We just did this trick using the example with flipping a coin five times. If A and B are two independent events, then the probability P(AB) that both events A and B will occur simultaneously is determined by the expression:

$$P(AB) = P(A) \times P(B).$$

For example, suppose an applicant to a college determined that his chances of admission are 40%. And the admission chances of his friend to another college are 20%. This means that the probability that both of them will get admitted to their respective colleges is $0.2 \times 0.4 = 0.08$ (or 8%).

In the case where A and B are dependent events, the probability of both events occurring simultaneously is determined by the expression:

$$P(AB) = P(B) \times P(A|B).$$

Here, P(A|B) denotes the conditional probability of event A given that event B has occurred too. For example, suppose a college applicant has determined that his chances of admission are 40%, and he knows that housing will be provided for only 50% of all accepted students. As a result, the chances of being admitted and receiving housing in a dormitory are determined as $0.4 \times 0.5 = 0.2$ (or 20%).

The expression mentioned above derives from Bayes' theorem, named after its creator — Thomas Bayes (1702 – 1761), an English mathematician and clergyman. In the Bayesian interpretation, probability is constructed as a reasonable expectation representing a state of knowledge, or as a quantitative assessment of personal belief or faith.

In the case of flipping a coin, we use the principle of symmetry and, therefore, we get $1/2 = 0.5$ for the probability of getting "heads" (or "tails"). For a die, the probability is 1/6 for any one side to land facing up. However, there are many cases where the principle of symmetry does not apply. And there are situations where we can't say anything at all about our expectations for the probability value.

Imagine we have no information about an event that is about to happen. All we know is that there are several possibilities. In this case, we may assume that all possibilities are equally likely. For people familiar with statistics, this is known as the non-informative priors, introduced by Pierre-Simon Laplace (1749 – 1827), a French mathematician. This is one of the oldest principles — the principle of indifference, which assigns equal probabilities to all possibilities.

For example, you can ask the following question: What is the probability that it will be cloudy at exactly noon in 10 cities, evenly distributed across the Earth. We have absolutely no idea where these cities are located, and what the climates are in these cities. Surely, there must be some answer to the question about the probability of such an event. This probability is definitely not 0, but also not 1. These two values are excluded, following the reasonable expectation that there is an atmosphere and clouds on Earth. Therefore, it is rational to assume that there are two possibilities for the weather in one city: either there are clouds in the sky, or there aren't. Hence, the derived probability equals 1/2 = 0.5 using the principle of equal likelihood for the two possibilities. Applying the multiplication principle, the final probability of cloudiness in all 10 cities at the same time will be $(0.5)^{10}$ = 0.00098. This is a small value, and it matches our intuition. It suggests that if there were 1000 parallel worlds similar to Earth, then there would be only one world with 10 cities covered by clouds.

But is this computed value for such a probability the correct answer in the strict scientific sense? Of course — not. To perform this calculation, science requires historical weather and climate data for each town. Nevertheless, our reasoning does reflect everyday intuition, which tells us that we are dealing with a rare event. We do not need to worry too much about possible inaccuracy in the calculation because the very fact of multiplying probabilities solves this problem for us and leads to a small value. By the way, the probability obtained for cloudiness in 10 cities is not that far from the truth. You can check this yourself using real-time weather forecasts.

Even when we do not have any information about the probabilities of events, we still use some knowledge when assigning equal probabilities to all possible outcomes. For example, we know that there are cloudy and sunny days on Earth. We do not speak of Venus (where it is always cloudy) or Mercury (where it is always sunny). Therefore, even non-informative priors (knowledge acquired before experience and independently from it) are often based on some initial information and some belief.

Thus, for calculating probabilities, we will often use principles of symmetry which reflect some information about the essence of phenomena. In other situations, when we know absolutely nothing about the probabilities of events, we will use non-informative priors, assuming that all possibilities are equally likely.

Perhaps this is all the mathematics we need to know to understand this book.

3.4 About probabilities and information

In chapter 3.1 Entropy, we touched on the problem of entropy. As we determined, not every pattern made of letters can be called information, although such constructions of characters can have low entropy (or any entropy number you like).

What is the probability of creating a meaningful sentence in general? Imagine a monkey playing with cards or pictures. Each picture has one uppercase letter from the alphabet. There is also a picture with a space. How can we find out if this monkey understands the English language? What is the probability that the monkey, by playing with these pictures and drawing exactly 6 pictures blindly, will form a sentence like "I LIKE"? Although this probability can be calculated analytically, I will not tire you with such a calculation. Chapter 20.2 Appendix shows the program code for this calculation. The resulting probability will be as follows:

$$3 \times 10^{-9}.$$

This is an incredibly small probability for everyday life. If the monkey is pulling out random cards in an attempt to write a sentence that makes sense, such as "I LIKE SNOW" used in chapter 3.1 Entropy, then such a probability would be close to 0 for any practical purpose.

The example with the monkey gives you some intuition about what we're dealing with when we talk about the random creation of information, if such a permutation of molecules were to occur in some environment. In terms of magnitude, the probability of obtaining a word with N letters, when there are a total of M letters to choose from, decreases as $1/N^M$. In mathematics, the function $1/N^M$ is one of the fastest decreasing functions with increasing values of N and M. It seems that nature itself is trying to give us this obvious hint using mathematics. The material world, which randomly considers all possibilities, is unable to create even a small sequence that can be perceived as information.

Now let's try to solve a more complex problem. The initial sentence will be "I LIKE". The final one — "I LIKE SNOW". We will randomly draw four letters, and then substitute these random letters into the initial sentence until we get the required letters to form "I

LIKE SNOW". The example in chapter <u>20.3 Appendix</u> shows the program code that calculates this probability. Our result

$$7.4 \times 10^{-9}$$

again indicates a very small likelihood of such an event to occur. This demonstrates that "mutation", that is, adding or substituting a random letter in order to obtain some useful information, leads to probabilities that are as small in magnitude as the original probability of obtaining the initial word randomly.

Of course, our example is quite simple compared to those that could demonstrate the formation of life. On the other hand, it doesn't even remotely resemble the amount of information found in a cell's DNA — the molecule that carries genetic information for the functioning of an organism.

3.5 Life and information

Cells — the building blocks of living beings — are incredibly complex factories for storing and processing vast amounts of information. Typically, the DNA molecules are located in the cell nucleus. The DNA are long macromolecules that ensure the storage and transmission of genetic information from generation to generation for the functioning of living organisms. Other long molecules, such as RNA, encode and read genetic information.

If we imagine that a cell is a complex device similar to a computer, then understanding what DNA is can be achieved by comparing it to a hard drive that contains directives for cell operation. The information encoded in DNA represents instructions for performing actions that animate the cells, and ultimately, the living beings themselves, consisting of many trillions of cells. For such functions to occur, the presence of a conscious mind is not required. However, for the instructions to be correctly understood as signals for specific actions, the presence of a mind that created these molecular mechanisms is unavoidable. At least for those who do not believe in the creative role of dead matter.

Not everyone agrees with this opinion. A category of scientists who believe that simple biological organisms arose spontaneously from non-living molecules will argue that no consciousness or intent is needed to explain biological information. For them, information can, in principle, arise randomly.

Information from nothing is a brief definition of the approach where intelligent intervention is not needed in the emergence of life. This reasoning usually goes as follows (Dawkins 2006):

"At some point, a special molecule formed by chance. It was a replicator molecule that was the first to reproduce itself. Thus, it gained an advantage over other molecules in the primordial environment. Over time, replicating molecules became increasingly complex".

From the perspective of the informational approach, such reasoning has problems. Even if we imagine that some random process created a long molecule capable of making copies of itself, this says absolutely nothing about how to create the simplest cell, where data is stored and then read by other molecules for any action. Indeed, in nature, mechanisms that replicate themselves can be found. However, self-cloning is not a sign of life. One can create computer algorithms that make copies of certain complex structures, albeit not randomly, but after certain rules are defined (Wolfram 2002). This has little relevance to life, even as simple as that of the most primitive cell.

Even the oldest known single-cell organisms are unfathomably complex. For instance, the cell of the simplest organism is not just a mechanism for cloning, but a molecular "machine" separated from the environment by a membrane. It contains specialized and coordinated parts between which informational flows circulate. It must be able to store information describing the sequence of processes within the cell, react to the environment, and have the ability to transmit it to its offspring (in an altered configuration!). We already know from the previous chapter that even the simplest information, in the form of directives for performing certain actions, cannot be obtained through random selection and the laws of inanimate matter.

One of the most precise definitions of life is given in the book "Cosmosapiens: Human Evolution from the Origin of the Universe" (Hands 2016). It sounds like this:

"Life is the ability of an enclosed entity to respond to changes within itself and in its environment, to extract energy and matter from its environment, and convert that energy and matter into internally directed activity that includes maintaining its own existence".

These processes, necessary to maintain such abilities, are deeply informational in nature. Note that the ability to produce offspring is not the main characteristic of life.

But first, a bit of biology to understand this quote. All cells are composed of proteins that perform specific functions. Some proteins function as tiny machines. Others act as building blocks. Each protein consists of a chain of amino acids, which are the building blocks of proteins. They link together into long chains, which ultimately fold into functional proteins. The simplest form of life consists of at least 250 proteins, and each protein consists (on average) of 350 amino acids. There are 20 essential amino acids that make up all life. But here arises some complexity. Amino acids exist in two forms: left-handed and right-handed. All living organisms consist of left-handed amino acids. If one right-handed amino acid gets into our amino acid chain, our protein will be destroyed.

There are various estimates of the probability of life originating spontaneously. To make such estimates, one needs to calculate the probability of creating one functional protein containing only 150 amino acids, assuming they appeared randomly. What is the probability of getting 150 left-handed amino acids in a row? Considering that the chances of obtaining a left-handed amino acid are 50%, the probability of getting 150 left-handed amino acids in a row is $(0.5)^{150}$, which equals 1 chance out of 10^{45}, or 1 followed by 45 zeros (Meyer 2009) (Barnett 2015). This is the same probability as flipping a coin 150 times in a row and getting heads or tails each time. Another estimate for obtaining a functional protein from all possible proteins is 1 in 10^{74} (Miller 2019).

For serious claims that life arose randomly somewhere in the Universe and ended up on Earth, where all of the conditions for the emergence of life were present, one must deal with astronomically small probabilities. For comparison, the number of subatomic particles in the Universe is 10^{80} (very approximate!).

Physicist Francis Crick (1916 – 2004), Nobel laureate in biology, stated in 1982:

"An honest man, armed with all the knowledge available to us now, can only state that in some sense, the origin of life appears at the moment to be almost a miracle, so many are the conditions which would have had to have been satisfied to get it going".

Molecular biologist James Watson and Francis Crick discovered DNA in the 1950s.

All we know now is that the spontaneous origin of life has not passed even the most elementary scientific tests. When it is said that life appeared on our planet spontaneously about half a billion years after the Earth formed, all you need to know is that there is not much evidence for such a hypothesis. Out of the millions of events observed on Earth now and in the past, humanity knows of no phenomenon or structure created spontaneously by inanimate nature without the involvement of a living entity that already carries information. Even the minimal amount of recorded information, such as the phrase "I AM HERE", requires a vast amount of random rearrangement in selecting letters, even for a modest number of letters in the alphabet. Useful information arises only from another source of information. And life arises only from life.

Let's move beyond the simplest cell and consider complex biological systems consisting of billions of cells. Each cell in our body encodes approximately 1.5 gigabytes of data. One gigabyte is about 10^9 bytes. To illustrate this, all of the data from the Library of Congress can easily be stored in a "DNA archive" the size of a poppy seed. All data ever created by humanity can be stored in a sphere made of DNA, which is no larger than a ping-pong ball (Lim 2021). An ordinary bacterium can store about 1.25 exabytes. One exabyte equals 10^{18} (quintillion) bytes. The average human body consists of 37.2 trillion cells. Thus, a body with 37.2 trillion cells would contain a staggering 55.8 billion terabytes of data. One terabyte equals 10^{12} bytes. We will not attempt to verify these calculations (Mushtaq 2023), but even if there is some margin of error in these estimates (since this issue is not fully understood), it is not very important. The scale of information in this example is just staggering.

Life, even in its smallest manifestation as a cell, is not just a database with information. The cell itself contains millions of molecular machines formed from organic molecules that use this information to construct other complex molecules, replicate this information, and animate the cell according to a given algorithm. The cell, in essence, is an incredibly complex information processing factory. It's not just a hard drive with information. These are molecular robots processing information and animating cellular processes.

Even if we assume that nature was able to create an organic molecule that began to combine with other molecules to create a single cell of an organism with 1.5 gigabytes and started processing it to animate all its parts, how could such a cell create organisms with information that was billions of times more useful? Here, I mean meaningful information that is useful for initiating processes leading to the increased complexity of microorganisms and the creation of the first complex organisms.

The simplest organisms formed and began transmitting their genetic information so that natural selection could take effect. It remains a mystery what triggered their transformation into super-complex conglomerates of cells, such as mammals, and ultimately, humans. The role of natural selection in local species changes is fairly well studied. However, it is still difficult to understand (Meyer 2013) why there was an increase in information useful for constructing an incredibly diverse array of species. Was it to enhance adaptation to the environment? Personally, I don't believe I am more adapted to the environment than any simple organism.

As I have already mentioned, to produce new information, a blueprint with some plan is needed. Without it, how can the simplest cells emerge? Even if you have some initial information, you cannot naturally generate new information that exceeds the initial by billions of times. You can either mix the initial data or randomly "infuse" in some new data elements (known in biology as mutations), and then sift out unsuccessful data combinations again. You will not gain any new information. This process is so insanely absurd that it is absolutely impossible to obtain a new, meaningful message. It is much easier to destroy an information record by a random process than to create, add, and then sift out the unnecessary. This is simply the law of information creation.

Here I completely agree with Stephen Meyer (Meyer 2013) (Meyer 2021), a scientist and a philosopher. From the perspective of the information approach, the formation of the simplest organisms through trial and error, random addition of information, and natural selection makes no sense. Intent and reason are the only economical ways to create life. In chapter 3.4 About probabilities and information, I provided the simplest numerical example illustrating what we are dealing with when discussing the emergence of complex information. The probability required for a chance to create the phrase "I LIKE SNOW" is astronomically small.

As a physicist, I know for sure that the only way to verify the validity of a hypothesis is to conduct an experiment. In my field (high-energy physics), there are dozens of quite elegant hypotheses that complement the Standard Model of physics. However, they have been refuted after their predictions were compared with experiments. If biologists set up an experiment that unequivocally proves the creation of complex organisms, such as those we observe now, starting from molecules or, at least, from the simplest cells, then I will gladly change my opinion. But we will come back to this later.

Perhaps it is worth remembering here that simple algorithms can indeed lead to complex structures using a few basic abstract rules (Wolfram 2002). Indeed, very complex behavior can arise from such rules. Intricate structures can result in very high (or low) entropy, as we show in chapter 3.1 Entropy. However, high or low entropy does not necessarily mean information capable of incredibly complex functionality. This is precisely what is required for the attributes of life. By and large, humans still do not know how to obtain meaningful information algorithmically or by influencing inorganic molecules. What are these "natural processes" that created the most complex informational product in the form of a cell from lifeless molecules? How can we seriously consider theories that operate on an incredibly distant past?

Biologists have no way out — they must work using a principle that leaves no room for design. As a result, they have to resort to all sorts of inventions to stay within the bounds of scientific methods. If you want to unsettle someone who wants to convince you that the first proto-cell appeared from chemical processes about four billion years ago, here are some questions you might ask:

- Can you provide an example of a natural process that creates a sequence of data that can be perceived by animals and humans as something meaningful or prompting certain actions?
- If a random process created a complicated formation, such as a colony of complex organic molecules, how can you prove that it will not disintegrate? After all, the number of adverse factors destroying complex molecules can be much greater than the favorable ones. This is what we observe in most cases. How do you refute this?

Scientists engaged in modeling, using complex scientific terminology, often forget how to return to simplicity. Such models or hypotheses may, at first glance, seem scientific. But upon closer examination, they are simply far-fetched fantasies, as the number of unfounded assumptions is so great that it is easier to explain everything with just one single definition — the presence of intelligent design. But this explanation does not adhere to the rules of science.

The emergence of information from the random processes of nature can be compared to the creation of a perpetual motion machine. The possibility of such a machine working indefinitely would mean obtaining energy from nothing. These devices may look complex and can sometimes be difficult to understand. However, the result of such contemplation is always the same — such engines are impossible, as their operation would contradict the laws of thermodynamics. Therefore, it is easier to approach the analysis of such machines by checking compliance with these laws rather than a detailed examination of the engineering design. The same applies to information — it cannot simply arise from the natural laws of the material world.

Some might say — well, we fully support the opinion that information cannot originate by itself. This means the question of the origin of life is definitively resolved! Life was created. This looks like a real theory!

However, this is not quite true. The claim that information cannot be created by matter can only be a well-argued hypothesis or an empirical law. It is very well supported via observations by all of humanity's historical experience. The incredible rarity of sequences perceived by humans as information is also known and verified by us in chapter 3.4 About probabilities and information using calculations. But this is not enough for a real theory, which must not only explain existing data but also have its predictions experimentally confirmed. Perhaps, this is the main problem for those trying to rigorously prove the existence of some force that created the world. We will return to the discussion of this topic at the end of this book.

The main feature in the hypotheses about the origin of life from molecules and its evolution into the most complex biological animals and humans is that they involve incredibly large spans of time — hundreds of millions and even billions of years. Using an analogy with mathematics, this is like finding a complex non-linear function from a few known points on the X-Y coordinate plane. In this case, such points are the fossil remains of ancient organisms, which are scattered in space and time. Since we cannot say with certainty how everything happened in the distant past, scientists fill the gaps in knowledge with assumptions. But how does one go about filling them? This is where these billions of years come to the rescue. One may say — anything can happen over such a period of time, as long as we know potential causes. And even life is capable of emerging from atoms and molecules if all of creativity is summoned to help.

As we have said before, life is not just recorded information inside a cell. Two more components are needed: An information creator and some agent that reads and understands the information. They are the ones animating molecular robots performing their functions in the cell. They must use the same protocol for encoding and decoding information. This is only possible when the whole system is conceived as a single entity. In such cases, an external agent who designed the entire system is needed.

Perhaps here I will stop, reproducing a point of view that is beginning to become popular about the emergence of life from inanimate matter beyond the bounds of the scientific method (Hands 2016).

3.6 Theory of evolution

The emergence of new animal species from existing ones is another big question that is still awaiting its answer. According to the original theory of evolution (Darwin 1859) by Charles Darwin (1809 – 1882), an English naturalist and traveler, the observed forms of life are not the result of the creative activity of an intelligent Creator, but variability, heredity, and natural selection. Natural selection is the main process driving evolution. It leaves only the strongest and most adaptable living organisms. As a result, the remaining strong individuals are capable of producing healthy offspring that can survive in changing environmental conditions.

The modern theory of evolution encompasses many components, such as natural selection, genetic variability, the formation of adaptations, speciation due to external factors, and so forth. It resembles more of a collection of various mechanisms aimed at demonstrating a common origin; that all living organisms originated from a single common ancestor through the process of natural development. To stay within the bounds of the scientific method, nothing extraneous beyond natural factors is assumed. Henceforth, we will simply refer to the modern theory of evolution as "Evolution Theory" (for brevity).

Can the collection of mechanisms in the Evolution Theory change the structure of living organisms and produce new complex plants and animals from other complex organisms? Why not? The modern evolutionary theory is supported by a plethora of evidence from various fields, including paleontology, comparative anatomy, molecular biology, geography and laboratory experiments on simple organisms.

However, the question I want to pose is this: Can evolution explain all the details of nature, with its diversity and the emergence of 8.7 million species of organisms, exactly as we see them now? Notice how this question is framed. I'm not asking whether the Evolution Theory provides a convincing explanation of why and how organisms can change. That question is already resolved for most scientists.

Evolutionary biologists would respond that the Theory of Evolution was specifically developed to explain living organisms. It is capable of addressing questions about all the details of their structure and how they came to be. This theory has been extensively tested, with a large number of experiments and observations. We can even track how species change by studying their genes.

Let's suppose that this is a theory indeed. It's up to biologists to decide what to call the collection of various natural and genetic mechanisms that alter organisms. But the question is still this: Can it provide a theoretical answer to the question I posed? Namely, how to explain everything we see around us, in all its details? The answer may be this: It is a theory, certainly, but it cannot provide detailed predictions of the surrounding world as we see it now since there is a lack of historical data on how the formation of a particular animal or plant occurred.

It is precisely at this point that you begin to understand that you are dealing with a belief system for such kinds of questions. After all, the Evolution Theory cannot answer this question. There is a belief that when we obtain a sufficient amount of historical data, evolutionary mechanisms will surely explain why organisms are the way we see them now. However, at the moment, such a theory cannot say anything specific about the question I posed. There are no data on precise explanations for specific circumstances. Therefore, we cannot test it on a historical scale, which is precisely required to understand the full diversity of the biological world. Biologists often use the epithet "imperfect theory", adding on that perfect theories do not exist.

Here I can draw an analogy with other areas of knowledge. If I know that there is a theory of electromagnetism, I will have no problem calculating predictions for any system of electrical conductors with known initial conditions. It is a theory precisely because it has a specific domain where it can provide precise predictions. This theory can be easily tested in laboratories. But if I ask whether this theory can predict the strength of electric currents in all electrical conductors in the world, including the atmosphere, what will the answer be? Of course, not! We do not know all the initial conditions and properties of electrical conductors worldwide. The theory exists, but the predictions do not. There is no additional data which is unrelated to this theory itself, but which is absolutely needed to make the prediction. Moreover, I do not exclude that this theory may not work for some natural phenomena or materials with which humans have not yet encountered. It will need to be improved for these purposes. So, although it is an excellent theory, it cannot answer some general questions. It simply cannot be tested on the scales required to answer some questions. This is precisely the problem of reverse engineering a huge layer of information of the past that is beyond the power of even the most ideal theory. Such theories can only create hypotheses and some models to describe the deep past.

Here is my point of view. Evolution is a well-tested set of mechanisms for explaining why and how organisms change. But it cannot provide answers to the questions of why organisms are exactly as we see them now, in all their details. "Theory" is the highest achievable level of confidence in scientific knowledge. Here, the word "highest" means highest in terms of understanding, not in comparison to other explanatory methods. A theory must not only explain but also predict. Can evolution say anything about an octopus, its emergence, appearance and lack of changes in its body over millions of years? Absolutely not. Or why male narwhals have tusks? Agreed, it's a childish question. But such childish questions about some animal peculiarities lead to very elusive explanations from biologists, which can hardly be called answers. For example, a scientific answer might be: "The tusk in male narwhals is necessary for attracting females". But why a tusk and not something else? We don't have data on all the circumstances of how this happened historically. Evolution lacks quantitative and qualitative certainty for many specific observations; it only has confidence that there is a set of natural mechanisms that can lead to changes in organisms.

Thinking about this, I stumbled upon a very similar point of view (Aczel 2015):

"While it is unscientific and ignorant to not recognize that evolution provides us with a powerful principle that often explains what we see in the biological realm, it is equally unjustified to assume that evolution is a "perfect theory" that explains everything. A theory that cannot provide superior predictions of future outcomes and phenomena is not a complete theory".

This is true, but even a perfect theory cannot provide the answers we are looking for.

Here I will provide several examples from other fields of studies. For instance, there is the Standard Model — a theoretical framework in physics describing the weak and strong electromagnetic interactions of all elementary particles. It is a model, not a theory, because it has about free twenty parameters that cannot be predicted but can be experimentally measured. The Theory of Evolution does not even "dream" about such challenges in describing the formation of complex life! There is the theory of electromagnetism, with its concepts of electric charge and electromagnetic fields. This is indeed a theory, with all its characteristics (with strict predictions and well proved through laboratory testing). As we mentioned, even this theory cannot provide an exact answer to many general questions, especially related to the past of the Universe. The theory exists, it perfectly works, but there is no full set of data for explaining many phenomena using this theory. There is also quantum field theory. This is also a theory. It provides a wonderful and precise description of nature when we know all the initial conditions to derive predictions. In chemistry, there is the theory of chemical bonding, describing how each pair of atoms in a molecule is held together. This list of examples is easy to continue. We can confirm such theories in laboratories. They give precise predictions for specific cases and known conditions. Using these theories, smartphones can be created, and particle accelerators can be built, costing billions of dollars. All of this will work beautifully. These theories are comprehensive and complete. They are not "imperfect". Yes, they may evolve over time and merge into other theories, but they already quantitatively and qualitatively describe everything we know within the limits of their applicability. There are cosmological theories describing the development of the Universe. Just like in the case of describing historic developments of living creatures, they deal with events from the distant past. We cannot measure the predictions of cosmologists in laboratories. However, such theoretical descriptions of events in our Universe are built on laws and theories that have been well tested experimentally on Earth. Notice that cosmologists do not describe the evolution of information-rich environments of the past; they deal with more predictable natural processes. Nevertheless, in cosmology, such theoretical descriptions are predominantly referred to as models rather than theories.

I am absolutely confident that evolution works, and complex living organisms can change, adapting to their environment and other mechanisms of the Theory of Evolution. But are we sure that everything we see around us originated from a primordial cell? How can we prove models that deal with billions of years worth of historical data on the development of living beings?

The increase in the complexity of an organism is the evidence for the emergence of new information. Indeed, there are some laboratory experiments that have shown how bacteria, fruit flies, and some other species of organisms adapt and change even over a short time of experimentation. But going from such experiments to a theory that explains how the simplest biological cell led to humans over billions of years of evolution is an infinitely huge step. It does not matter that there are many scientists who believe that the Theory of Evolution is exactly a theory, rather than some framework uniting various known natural evolutionary mechanisms. Many scientists consider it the best existing explanation. Indeed, laboratory experiments can demonstrate how flies and bacteria adapt. Fossil remains have a similar structure, which can also speak of common ancestors. But this is not enough to prove the emergence of all the complexity of the animal world from simple cellular organisms, when we talk about billions of years of history (Meyer 2013).

From the perspective of computer modeling, work on cellular automata[1] has shown that incredible complexity can be achieved without natural selection (Wolfram 2002). Moreover, such modeling has been able to reproduce some properties of biological systems. All that is needed is to set a simple rule of behavior for such algorithms, and that all complexity arises without any selection and struggle for survival. Of course, who sets such rules in nature is a big question.

Theories, such as the Theory of Evolution, attempt to explain events over astronomical spans of time. They do not take into account the simple fact that billions of years (even millions of years) are an incredibly long historic period for information processes. It is impossible to imagine, nor to model on a computer. Time in such discussions is a kind of "miracle" used to explain everything, even an incredibly complex biological system with a huge amount of embedded information. But to this argument, there is always the following counterargument: Over such a large span of time, we simply do not know what can happen. Since we are talking about an immense span of time, anything can happen, even if the probability of such events is very small. We remember the law of large numbers — if you wait long enough,

---

[1] A cellular automata is a discrete computational model in which the state of a cell changes depending on the cells surrounding it.

something can always happen, because the number of possibilities increases. See chapter 3.3 About probabilities.

The role of random events and external circumstances potentially occurring over hundreds of millions of years is so significant that it is incredibly difficult, if even impossible, to construct a true theory with strong explanation power. These processes can be envisioned as massive peaks of sharp development (or degradation) occurring over a very short span of time. They represent sharp fluctuations in the behavior and organization of complex biological systems. These are informational "explosions". Such unpredictable and chaotic phenomena can lead to fundamental changes in DNA and to additional "injections" of information, leading to revolutionary changes in a very short period of time.

These scenarios are shown in Figure. 3.6. If you have several points on a surface, you can always extrapolate these points by some curved lines and claim that such a curve is the description of the data. This smooth line may be our conceived model. But the assertion that this curve line is indeed reality requires very serious evidence. Similarly, biologists studying evolution assume that the entire picture of species origin can be established from scattered fossilized remains. Just imagine billions of years of evolution, where small changes work over incredibly long-time scales — and an answer is obtained. But is it really the answer?

A typical example might be the Cambrian explosion, which occurred 541 million years ago at the start of the Paleozoic era. It is characterized as an unprecedentedly rapid emergence of diverse types of organisms. At that time, the major groups of multicellular animals appeared, many of which exist to this day. The Cambrian explosion occurred over tens of millions of years. Even over such a time span, the role of evolution was not minor. By the explosive appearance of information, I mean much shorter time intervals, such as decades or hundreds of years.

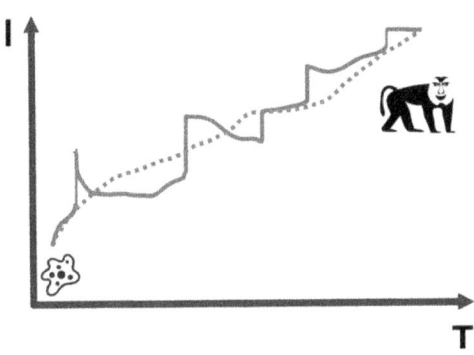

Figure. 3.6. *A schematic graph showing the growth of biological information, denoted by the letter "I", as a function of time ("T"). The dashed curve is our expectations according to the Theory of Evolution. The solid line with sharp jumps is one of the possible scenarios for the emergence of new information. Sudden jumps can occur very quickly, so on the full scale of billions of years, the thickness of the vertical lines would be imperceptible.*

Are there examples of such external and unpredictable phenomena? Of course. Simple organisms could have been present in meteorites. The presence of microorganisms in meteorite fragments has already been proven. Another example is that the Earth could have been seeded with the simplest life during a massive bombardment by comets. Given that there is a lot of information about unidentified flying objects, complex life could have been "edited" by intelligent beings from other planets. Sudden volcanic activity could expose a colony of living organisms to radiation, leading to rapid mutation, while destroying all other organisms that had accumulated evolutionary advantages. Or simply a herd of animals could have fallen into a gorge with completely different conditions, leading to rapid mutation within a few generations. The Theory of Evolution explains that these circumstances are already included in its mechanisms of evolution. This is good, but do we know how they acted in particular situations in the past?

These examples could go on endlessly. Hundreds of millions of years should lead to a vast number of unlikely events that could have a huge impact on the informational component of life and its changes over time. Furthermore, we will show that even over the course of hundreds of years, or even within the span of one person's lifetime, inexplicable phenomena and circumstances leading to radical changes are quite possible, not to mention billions of years worth of changes.

Modern models that explain complexity from simple to complex suggest that there is a gradual accumulation of changes; one just needs to imagine how they slowly accumulate from the beginning of life's formation. Unfortunately, it is impossible to see how this can happen. Archaeologists and biologists work with small slices of time, studying the remains of animals in the geological layers of the earth. The time span between these geological strata is tens or hundreds of millions of years. The archaeological dig sites themselves are spread out over vast distances on the Earth's surface. Mostly, we get information from fossils. They arise under very specific conditions — when the voids of soft tissues are filled with groundwater, followed by mineralization. Such a process must occur exceptionally quickly. Therefore, in the history of the Earth, such events occur very rarely, as a combination of many rare circumstances. Information about the overwhelming majority of ancient inhabitants does not reach us — they disintegrate before they can even fossilize. Therefore, it is impossible to create a theory capable of building a continuous description and accounting for short peaks in the emergence of new biological information.

The evolution of living beings somewhat resembles the evolution of societies. In the mid-19th century, English philosopher Herbert Spencer (1820 – 1903) proposed that society is an evolving organism, similar to a living organism studied by biological science. The term "evolution" itself is Spencer's. According to him, society grows, increases in its volume (population), and becomes more complex. Then begins the division of functions. The informational component of society increases. It becomes similar to a complex organism. Like biological organisms, society adapts to the environment. It can "devour" more primitive societies. Like the brains of animals, in social organisms, there are governments responsible for decision-making.

Can the development of societies be described by pure evolution and slowly occurring mutations, as taught by biology? In part. In the history of societies, random and unpredictable events have an incredibly huge significance. They occur in very short periods of time, but strongly determine the further evolution of society. Randomness in history, as in nature, "acts" alongside necessity. Many scientists have now come to this conclusion. For example, if Cleopatra's nose had been a little shorter or longer, she would not have been so beautiful, and perhaps would not have had so many relationships with various Roman generals. This would have changed the course of history for many societies in the territory of the Roman Empire. As Blaise Pascal (1623 – 1662), the French mathematician and philosopher wrote: "Cleopatra's nose, had it been shorter, the whole face of the world would have been changed". The European invasion of the American continents was an accidental occurrence for the American civilizations existing at that time. In 1185, warriors of Prince Igor, who set out on a campaign against the Cumans, observed a solar eclipse.

This random coincidence of a solar eclipse before the start of the campaign was taken by many as an ominous omen. The boyars accompanying Prince Igor on the campaign tried to explain to the prince that they should turn back, as this sign did not bode well for them. The prince did not listen. Later on, Igor and many princes were taken prisoner by the Cumans, and many warriors died. It is well possible that this eclipse may have wielded significant psychological influence in the Russian army's defeat (Балашов 2022). Generally, the main reasons for the significant changes in the history of Russia were unpredicted events, as the Soviet and Russian philosopher S.A. Ekshut believes (Ekshut 1994).

As we can see, chance and unpredictable events in history play a huge role. With only archaeological finds, stone tombs of kings and ruins of castles, and without written archival records, we would never understand the details of the paths of societal development. Of course, we could guess at some evolution of societies, similar to the guesses of Spencer, but it would not be a complete picture in explaining the processes of societal development.

Scientists, creating models of the past, act like forensic experts, using mineral and organic formations of the earth's crust. Unfortunately, there are no witnesses to what happened in the past. This is an important point for information-defined events. Let me present an analogy here: Suppose you need to remember what you ate 3 days ago. You remember nothing about that time. There are no records or witnesses. All you can do is rummage through the trash and find scraps of food packaging or check a receipt from the store. Even if you find eggshells and suppose the time when you threw them out, it does not at all mean that you ate scrambled eggs three days ago. This could be a good hypothesis if you know that you cannot obtain any more information about the past. However, many other circumstances are also possible! What if a friend came over and prepared their favorite egg dish, which you've never eaten before? Maybe you baked a cake? Or you decided to cook a new dish after seeing an advertisement on TV? As we can see, even in such a simple situation, you cannot be certain about anything. Knowing this, how can one discuss scales of time that a person cannot even imagine? Therefore, all information about historical events and how something happened over billions of years are merely models and hypotheses.

In my view, an imperfect theory can succinctly be defined as a model. Such a concept as an imperfect theory "asks" for the presence of other epithets like "just a theory", "perfect theory", and so on. Everything in the scientific method consists of hypotheses, models, empirical laws and theories. Darwin's hypothesis from the mid-18th century became a model, or an empirical law, after undergoing certain checks in the 19th century that proved its validity. Models become a scientific theory after various independent experiments, and when they start having high explanatory and predictive power. This is how we understand the word "theory" in natural sciences. The Theory of Evolution is a theoretical framework that unites natural mechanisms and models to explain organismal changes. When dealing with historical information-rich events that cannot be observed or reproduced experimentally, such mechanisms cannot explain all the richness of specific organism features. This has nothing to do with the mechanisms united under the roof of this theory. The missing data to acquire reliable predictions is the problem. We lack information about the sequence in which all evolutionary processes occurred, when some processes were activated while other mechanisms of evolution ceased to function, or perhaps if they were acting simultaneously.

The Theory of Evolution is unable to answer a long list of questions it faces in explaining the vast diversity of biological forms around us. Why are there animals whose ancestors have not yet been found? Such "living fossils" as sharks, crocodiles, and some species of crabs have existed for vast periods of time. They have not under-gone evolution over many tens of millions of years, although other animals have evolved. Trilobites[2] suddenly appeared in the Cambrian period, about 520 million years ago, and were the most advanced forms of life on Earth. But we do not know how they appeared in such a short period of time. Their primitive ancestors have not been discov-ered. We also do not know what factors determined the complex struc-ture of trilobites, their appearance, and behavior. We have not found the "cradle" of such early experiments with complex life.

The Theory of Evolution cannot be tested for the vast majority of animal species surrounding us. What prediction from the Theory of Evolution do you know that has been confirmed over time? I have been fortunate to learn about confirmation of such predictions for sev-eral simple organisms. But there are approximately 8.7 million species on Earth. Of course, one can imagine that in the future, we will find all the explanations and prove the predictive ability of evolution to create complex organisms from simple cells. Then we can say that the Theory of Evolution is indeed the theory to solve such questions. But for now, this theory simply serves as a unifying principle or a frame-work for various scientifically grounded mechanisms that explain the variability of complex organisms, without precise quantitative and qualitative predictions of our past.

I once heard that articles confirming the Theory of Evolution are often not accepted for publication in journals, as they contain noth-ing new, and that this theory itself does not require further proof. What could be more ironic?

---

[2] Trilobites were marine arthropods that became extinct over 200 million years ago. They belong to the same group as spiders, insects, and crabs. They could reach up to 70 centimeters in length.

3.7 Brain and information

As we have already shown, consciousness is the primary cause of information, its creator, and its consumer. Intelligence is used as a supplementary tool to aid in the creation, processing, and consumption of information for certain actions. However, all such operations of intelligence occur when the decision is made by consciousness. It is consciousness that decides how to interpret the results of such calculations. To create and perceive information, one must abstract away from the concepts of this world. One must realize oneself as something other than matter and time. One must "rise" above matter and separate oneself from it. This is what consciousness does. And by realizing yourself, you gain a sense of meaning and feeling of what you are. If you know that this world does not belong to you, then it means you have a reason to be separated from the world. But what is this reason?

Only by separating yourself from something, do you get the opportunity to modify the environment from which you have detached yourself. By observing and understanding the external environment, you gain the ability to create something new in it. The ability of consciousness to create objects using intelligence and its biological body is the main feature of a human being. Creation can be anything. From a machine to the Universe.

In materialistic understanding, the brain is the generator of consciousness — the entirety of thoughts, ideas, sensual and mental images that allow a person to realize the fact of their own existence. This organ exists for the purpose of processing and storing information. It also creates new information. Consciousness and the sense of "I" are by-products of the brain's activity.

71

Another influential direction in philosophy is dualism. It recognizes the equality of two principles that are irreducible to each other — spirit and matter, the ideal and the material. A large number of philosophers who have had a tremendous influence on society were dualists, such as Plato, Aristotle and Descartes. Dualists assert that the mind and the brain are not the same. Dualists and materialists have been at odds for centuries. However, the ultimate answer may turn out to be a surprise for everyone (see chapter <u>3.9 Idealism and information</u>).

The human brain is incredibly complex. The brain of an adult human has about 100 billion neurons, each of which is connected to more than 1,000 other neurons. Neurons communicate with each other using electrical impulses. The human cerebral cortex is the largest among mammals in terms of its relative size, making up more than 80% of the brain's mass. However, other primates (as well as non-primates) also have quite impressive brain sizes. This concerns both the ratio of brain weight to body weight and the absolute size of the brain. For example, the number of neurons in the brain of a whale is 5 times larger than that of a human.

At the other end of the scale of living creatures, an ant's brain contains only about 250,000 neurons. Ants have developed efficient methods of agriculture, mastered animal husbandry, navigation, invented slavery, and established a caste system. Individually, ants are quite simple-minded. However, within anthills, they exhibit an astonishing complexity of organization. This is because the anthill is akin to an organism with a brain distributed among many ants. Ant nests resemble a collective intelligence. Each ant is a separate mobile processor, or a small autonomous piece of a brain that moves and exchanges signals. An anthill with 1 million worker ants can be considered a huge "superbrain" with 250 billion neurons. The connections between such autonomous pieces of brain (ants) occur in various ways. The primary means of communication among ants are pheromones, bioactive substances released by ants.

The discrepancy between our remarkable intellectual abilities and our not-so-impressive brain size suggests that the human brain is an exception to the rules of the animal kingdom. Is there consciousness in animals? Almost certainly. The primary method of detecting consciousness is the mirror test, which reveals the ability to recognize oneself in a mirror and distinguish oneself from others. Primates and dolphins easily pass this test. Domestic animals, such as cats and dogs, do not pass it. Even children up to 15 months old (on average) are not able to recognize themselves in a mirror. This does not mean they lack consciousness. They definitely have self-awareness. It's just that animals rely more on instincts, rather than intellect. They don't need a developed cerebral cortex because the instinctive memory of how to react to the external environment doesn't require large mental resources. Therefore, there is no need for a developed cerebral cortex. In nature, instincts work much more efficiently and react faster to danger. One could even imagine a fantastic scenario: If one's consciousness is an alien from the immaterial world of ideas and forms, then seeing one's reflection in a mirror is not for the faint-hearted. It takes time to get used to and understand that it is possible to have a body in this reality and, therefore, see it in objects that reflect light.

Consciousness and intellectual abilities depend on each other. Some believe that consciousness emerges at a very advanced level of intellect. The greater the computational capabilities of the intellect and the flow of sensory information, the higher the degree of consciousness. In this case, computers will acquire consciousness at some stage of the development of artificial intelligence technologies. Therefore, someday, a certain robot sweeping a floor somewhere will approach a mirror, recognize itself, and ask, "Is this me? But why?".

Others, and I am among them, consider consciousness to be primary. Intelligence, as the computational part of the mind, only serves consciousness. The greater the degree of self-awareness, the more "power" of intelligence is needed to serve the presence of consciousness in the body. For quick reactions to the dangers of the surrounding world, complex instincts are necessary because intellect is a much slower way to react. More active use of instincts and a lesser role of intellect do not always mean less self-awareness.

Scientists do not have a definitive answer to how consciousness and intellect relate to each other. But those who have pets or have had experience interacting with them will almost certainly agree with the point of view that consciousness is an inherent feature of animals. They have long "voted" for this position, without waiting for scientific proof. For them, animals have consciousness or a soul.

Sophisticated intelligence, despite its drawback of being slower, holds distinct advantages: Unlike instincts, it offers greater flexibility and adaptability to changing circumstances. Animals, including humans, can navigate unconventional situations throughout their lives, a feat more challenging for those reliant solely on instinct.

Relying on instincts for survival in the natural environment, animals are often perceived as something inferior. However, their way of interacting with the surrounding world of atoms and molecules is simply arranged differently. Perhaps they are the "alien mind" we are looking for. They have been on Earth much earlier than us. However, "brethren of consciousness" might be a more fitting term for animals. Mind includes intellect, but animals do not need great necessity for it. Almost certainly, most beings on other planets will also use instincts to maintain the presence of consciousness in the world of molecules. Maybe this is the reason for the absence of contact with extraterrestrial intelligence. Intelligence is an exceptionally rare gift. It is not at all necessary for life. Consciences in the material realm do not always need it. The fact that humans are endowed with it makes them special on a cosmic scale.

And if consciousness is what intellect and instincts form around, then computers will never be able to become self-aware, no matter how complex they are. The question "Is this me? But why?" posed by an advanced robot in front of a mirror would simply be programmed into some algorithm or generated because of a random question generator in its software. We have already said earlier that calculations in computers are created by program codes, which are simply "frozen casts" of the human mind. They can only run algorithms and appear intelligent.

It may be incredible, but science has still not determined whether the human brain is a generator or receiver of consciousness. Solving this dilemma is incredibly difficult. The answer to this question depends on one's views and belief system.

We know that brain damage disrupts its computational functions. This is exactly what happens with computers in case of malfunction. But the same can be said about a device receiving signals. When brain functions are disrupted, the signal may be altered, some information may be lost, or we may not receive the signal at all due to the brain being tuned to the wrong "wave".

Suppose you have a gadget on your desk — Amazon Echo or Google Home. You ask a question — it answers. Such devices indeed perform some calculations in real-time and collect information. But as soon as the task to be solved becomes complex, such devices send a request to a very powerful main server and receive the answer over the internet. In this case, these devices are receivers. It's completely unreasonable to have a supercomputer on your desk that would perform complex calculations and find the right answer from a huge database.

In biology, there is one thing that can be said with absolute certainty: The structure of organisms is always optimized for life activity. The fact that our brain can receive and process information, perform certain calculations in real-time, and simultaneously function as a receiver and transmitter could be a reasonable hypothesis.

And if that's the case, then this opens up an incredible number of phenomena that are impossible to explain from the perspective that we are robots with a computer in the upper part of the body — the head. The information received by the brain may have a single source. And this means, perhaps, that we all have access to this source. Thus, we are all connected through it.

There is no scientific evidence that the brain can perform functions that do not fit into the concept of conventional computing, as in computers. But there are some hints. One of the most mysterious properties of the brain can be discovered in cases of severe brain damage without the loss of physical and mental abilities. For example, in 2007, a 44-year-old man was found in France with only 10% of his brain tissue remaining (Feuillet, Dufour, and Pelletier 2007). The rest of the skull was filled with cerebrospinal fluid. Despite this, the man was capable of living a normal life, working, and having a family. Like all people, he felt an "I" within himself. The intelligence quotient (also known as IQ score) was equal to 75, which corresponds to approximately 20% of people. But in the man's brain, the frontal, parietal, temporal, and occipital lobes were almost completely missing. One of the assumptions why intellectual abilities were minimally affected was the brain's ability to adapt and transfer intellectual functions to the remaining small parts of brain tissue. Maybe the brain is capable of recalibrating. Or is there something else?

Experiments with animals also deserve attention. They showed that changes in the brain do not lead to any major changes in their behavior (Pietsch 1981). In such experiments, salamanders were used. During the experiments, parts of the salamanders' brains were removed or altered. Their brains were even flipped. However, the salamanders were still able to function relatively normally. This led to the assumption that the brain could work like a hologram, where memory is distributed everywhere, and each part contains the whole in a blurred form, allowing many functions to be restored.

It is quite obvious that this ability of the brain is the complete opposite of what happens in computers. A minor malfunction in the central processor and memory unit inevitably leads to a catastrophic failure of the computer.

The fact that the brain can easily restore intellectual abilities after damage serves as some confirmation that not all calculations may occur in the brain. If you have ever been involved in radio crafting, then you probably know that there are many different types of radio receivers, from very simple to complex. The simplest radio receiver does not have amplifying elements and does not require a power source. Such "detector" radios can operate receiving power only from the energy of radio waves. They contain only a few parts — a coil of wire, a variable capacitor, an antenna, a diode, and headphones. There are more complex radio receivers in which electrical signals are captured and amplified. They contain transistors. There are receivers with multiple blocks for amplifying the signal and variable resistors for tuning to waves, changing volume, and adjusting timbre. There are receivers with digital settings for the frequency of waves and for remembering stations.

When one damages some part of a sophisticated radio receiver, the effect of the malfunction will be somewhat different than in the case of a computer. After damaging one part of the receiver, you still have a good chance of hearing a radio station. For example, in the event of a breakdown, you may not be able to adjust the volume, or you will start hearing interference, or the station switch will stop working. One of the lessons I remembered from childhood assembling radios was this: Once, having assembled a transistor radio, I made a mistake. The part of the circuit where signal amplification was supposed to occur stopped working. One of the transistors burned out. But, by some miracle, I could still hear a faint radio station. Apparently, the energy of the radio wave was sufficient for signal reception.

As you can see, there is some analogy here with the brain. Devices that receive signals can still function even with some damage because all the work of creating information is located outside such devices. The purpose of the receiver is to find the signal and amplify it. Numerous components in a modern radio serve the sole purpose of enhancing signal reception, thereby making it more convenient for us to tune in and listen. Perhaps the purpose of the brain is also this, plus some other functions necessary for the body itself to operate. Of course, this is a very intriguing hypothesis.

As we said before, the main property of consciousness is the ability to distinguish itself from the surrounding environment as something special. This is completely different from the functions of intelligence — the ability to analyze, compute and solve tasks. It is consciousness that makes decisions because it contains free will. As we have said before, the notion that consciousness is a process, not something existing objectively, is a basic hypothesis in the material understanding of the world. Indeed, it is a process — you cannot touch it, just as you cannot touch music. However, if it is a process, then any process proceeds according to certain laws. For example, the process of nuclear fission occurs according to laws that we can discover and record. One of the popular explanations for the laws of nature is that they are not subjective. Science merely discovers them. The laws are set before the formation of matter itself and underlie all processes.

Matter cannot create the laws by which it operates. See chapter 7.2 The Lawgiver. Therefore, the consciousness process must have been created before an incredibly complex system, being our brain, appeared.

Secondly, consciousness is a very unusual process. It can switch from one type of activity to another, depending on its decision. All processes in the physical world do not change from case to case. They are strictly fixed and cannot change at their discretion, unlike consciousness. Everything that has ever been created by humans and nature has always had specific functions. For example, body organs or complex machines can only perform certain functions that are initially set. They cannot change such functions by themselves over time. This is very different from the brain, which can perform completely different functions depending on its desires. Feeling, writing poetry, experimenting, loving, and so on. If consciousness is a subjective process of the brain, like the music played by a record player, then it cannot change what it does at every moment of its life. Thus, consciousness is a completely new process, not belonging to the world of matter. Perhaps, this process was planned before brain cells began to perform such actions. The principles of this process were defined from the outside. Consciousness, like a certain algorithm, was planned outside matter, just like the laws of nature themselves, which must have existed before time, space, and matter appeared.

In this case, this world is not our consciousness's home. It carries the memory of an environment where neither matter, space, nor time existed. And that is precisely why, unlike the most sophisticated computer, we are capable of abstract perception and description of this world. Humans are capable of asking questions which cannot spontaneously appear inside a machine. We can separate ourselves from matter and can receive an awareness of our presence.

The reader may ask: Where is the evidence that our mind has direct contact with the world of ideals? Where are the particles or fields that facilitate such interaction? Of course, there is no direct laboratory proof that such interactions take place. That's precisely why such a hypothesis is speculative. However, we operate within the framework of an idealistic worldview (see chapter 3.9 Idealism and information). It asserts that energy, matter, space and time are secondary. Figuratively speaking, the material world is like a movie played by a projector that is not part of the movie itself. Therefore, such a projector is capable not only of creating but also of contemplating the entire spectacle from the outside. It is this feature of being separate from matter that makes consciousness something special and creates the sensation of an inner "self". From this, it follows that our consciousness's access to another informational reality does not occur through physical fields or particles. It is part of the information fabric of the Universe itself, on which all actions take place.

As Pablo Picasso (1881 – 1973), a Spanish artist and the founder of Cubism, said: *"Computers are useless. They can only give you answers"*. Indeed, a computer can ask a question if you program this question into it. But it will be your question, not the computer's. For example, we can ask — "Who am I?". This is not a question created by the machine or algorithm. It arises suddenly, at some random moment, inadvertently. And it is at such moments that we realize that all of this, including us, somehow exists in this Universe. We can observe all of this, analyze, and marvel that we are here. Such questions can only arise in a consciousness that has found itself in an unusual reality, locked in a garment made of chemical molecules, made for interacting with an alien environment. And some of us, who can come to this understanding, may even conclude that this world is not the true homeland of our spirit.

3.8 Will science explain everything?

Often you can hear such a phrase: "Since science has success-fully coped with explaining the mechanisms of the world's phenom-ena, it will answer all the questions in the future that currently have no answers". Perhaps, this is the biggest misconception out there. We have no evidence to claim that science can explain everything. All we have are examples where science has explained one phenomenon by another phenomenon. Science is a special way of logically describing and understanding the world, based on empirical verification. We can-not apply science to prove the statement "science can explain every-thing". We need to use something else. But what?

Science works effectively in situations where a phenomenon can be observed and an experiment can be conducted, isolating the problem from the random influence of many external factors. The ex-periment should be easily verifiable by independent researchers, and easily repeatable. When measurements and experiments cannot be conducted, science can only make assumptions and build models. The fact that there is a set of historical examples when science was suc-cessful in explaining something cannot prove the assertion "science can explain everything". Whether to believe or not in such an expla-nation depends on your worldview.

Science may never provide definitive answers to questions about the origins of the Universe, life and humans, as the exact cir-cumstances of their emergence are obscured by the deep past. They are not observable today. Not a single experiment is possible. All we can do is build consistent hypotheses and models.

But there are other types of questions that science fundamen-tally cannot answer. The British mathematician John Lennox provided a fitting analogy. Suppose you see a boiling kettle of water and ask – why is the water boiling? A scientist will explain this phenomenon in

81

terms of physical laws – temperature and external air pressure, etc. But another correct answer is – "because somebody wants to drink tea". This is also the correct answer, but it comes from a different level of understanding and another perspective on the issue. Perhaps, most people expect exactly such an explanation when they see a boiling kettle.

Such answers to questions about purpose suggest that complex things and phenomena may have some meaning in their occurrence. It is quite obvious that much of what we see around us originates from some other phenomena. These phenomena were the real cause. These mechanisms of the emergence of something from other phenomena or things can be revealed by science. But if there is some movement towards a certain goal, then this movement itself may arise because of some purpose. The laws by which such movement occurs may also arise from some initial idea or plan. It is utterly unreasonable to limit oneself only to questions explaining the mechanics and structure of something from other natural phenomena.

3.9 Idealism and information

This book would be incomplete if we did not discuss idealism — one of the oldest philosophical concepts. Idealism places ideas and meaning first. It postulates that the foundation of all existing things are ideas, consciousness, and the spiritual world, not the material world.

The opposite in meaning is materialism. This is a doctrine where matter and energy play the primary role. In this approach, life and consciousness are something secondary, which arose as a result of unconscious physical and chemical processes. Inanimate chemical elements, somewhere far in the depths of space and time, "decided" to come together and create the first primitive cell. Since it could reproduce, this cell produced something incredibly complex — a microorganism. I use quotes here because inanimate matter decides nothing by itself. This happened spontaneously, as a result of the infinitely long random permutations of molecules in the vast space. Then, this cell or proto-microorganism reached Earth, overcoming an incredibly long path, and was subjected to the incredibly harsh conditions of cosmic radiation and cold. And thus, life appeared on Earth. Alternatively, it can be assumed that the first cell emerged by itself right on Earth. For this, a favorable mix of chemical substances filling the oceans or geysers of young Earth was necessary. Somehow, amino acids formed randomly, leading to the emergence of the first protein compounds, which created more complex nucleic acids. As a result, the first cell capable of reproducing emerged, followed by multicellular organisms.

We have already considered the problems relating to the formation of life in chapter 3.5 Life and information. The problem of life is the problem of the spontaneous formation of information from matter. Out of the entire body of knowledge that humanity possesses, the probability of the emergence of a primitive cell from inanimate matter is astronomically small.

In essence, materialism suggests that the Universe came about by some natural method, or it has always existed, or it appeared from nothing as a result of the Big Bang. But all these processes were unconscious. Then, life appeared as the culmination of the gradual creation of new information by random interactions. This is the process of spontaneous and unconscious creation of functional complexity and information out of nothing. If it looks like magic, then the materialist's response is simple: "Just wait, and science will explain everything. Someday". This is precisely the same assertion that lacks scientific proof, as we explained in the previous chapter. This confidence is an element of faith and personal worldview.

On the contrary, idealism posits the primacy of consciousness, thought and spiritual essence, beyond and independent of human consciousness. Our world of matter and energy is considered secondary. At the foundation of the world lie only ideas and spirit, which do not adhere to the laws of the material world.

Idealism traces its roots deep into the centuries, to ancient philosophy. In Plato's idealism (5th – 4th centuries BCE), the spirit was considered as the formative element of inanimate matter. Plato argued that the true reality is a special, supersensory space of ideas. The material world is a space of shadows and pale reflections of the world of abstract forms, ideas and truth.

The French philosopher René Descartes (1596 – 1650) was one of the first to declare, *"All that we truly know is what is in our consciousness"*. The entire external world is simply an idea or an image in our consciousness.

Later on, philosophers of the Modern Era — Gottfried Leibniz (1646 – 1716) and Georg Hegel (1770 – 1831) — developed the idea of objective idealism. They acknowledged the existence of a universal soul that generated everything and dominates all material things. Thanks to this spiritual principle, humans are capable of understanding abstract and immaterial phenomena, guiding themselves with ethics and morality, and experiencing profound feelings, such as love.

Gottfried Leibniz founded a form of idealism known as panpsychism. He believed that the true atoms of the Universe are "monads". They are individual, non-interacting "substantial forms of being" that possess perception. For Leibniz, the external world is a spiritual phenomenon which is the result of a dynamic force dependent on these simple and immaterial monads. The central monad is God. He created a pre-established harmony between the inner world in the consciousness of waking monads and the external world of real objects. The world, in essence, is an idea in the perception of monads.

Other philosophers, such as Bishop George Berkeley (1685 – 1753), David Hume (1711 – 1786) and Immanuel Kant (1724 – 1804) established "subjective idealism". In this concept, the objects around us are nothing more than derivatives of our senses. For example, for Berkeley, material objects are simply ideas obtained through perceptual activity, and their attributes are sensory rather than physical properties. Sensation is impossible without the presence of ideas. All physical things in this world are divine ideas and are defined as "to be is to be perceived". And all people are ideas in the mind of God. When he thinks about us, we are born, and our existence is activated.

Kant's idealism is known as transcendental idealism. Transcendental idealism is the viewpoint according to which our perception of things is connected to how they appear to us (representations), and not with things as they are in themselves. There is a supersensible reality beyond the categories of human reason, which he called "noumenon". It roughly translates as "thing-in-itself". In this framework, Kant proposed that our knowledge of the external world is shaped not by the external world itself, but by the way our mind structures our experiences. According to Kant, the mind has a set of innate categories, such as causality, space and time, which it applies to the raw data of sensation to produce the world of experience. These categories do not apply to things as they are in themselves, but only to our experience of things.

I cannot fail to mention the scheme of the British theoretical physicist Roger Penrose. To understand why abstract concepts invented by humans make sense in describing the world, he popularized the concept of "Three Worlds": the world of Platonic forms, the Physical world and the Mental world (Penrose 2007). Figure 3.9 reproduces his model of reality. Such a scheme arises quite naturally due to the incredible property of abstract mathematics to describe the observable world. Note that only a small part of each world is projected into the other. Mathematics, related to the abstract world of ideas and mentality (or consciousness), must be outside the Physical world. We will return to the question of mathematics in chapter 9 Coincidences in numbers.

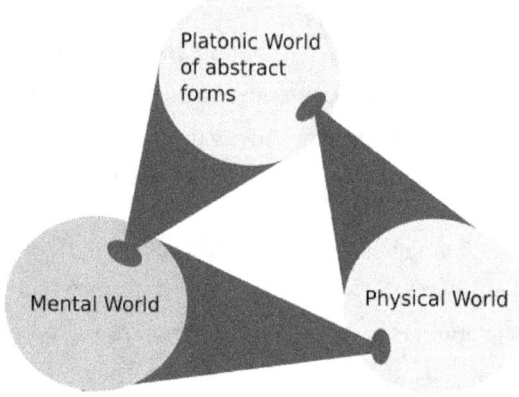

Figure. 3.9. *The "Three Worlds" of R. Penrose (Penrose 2007): The world of Platonic abstract forms, the Physical world and the Mental world. The world of abstract forms includes concepts such as beauty and morality. Only a small part of each world is accessible to the other worlds.*

The concept of the world of abstract forms, ideas and truth can be found in almost every religion. Hindus believe that we all are part of Brahman — the true reality of things. In Buddhism, nirvana has a similar meaning — it is the realization of one's soul, achieved through renunciation of material goods, comfortable external conditions, unnecessary desires and attachments.

I will no longer tire you with the recounting of the various concepts of idealism. I will simply point out that any such concept, in one way or another, asserts that meaningful transcendent[3] information is the primary source of reality. Ultimately, idealism and dualism lead to the idea of God. But God is usually associated with religion, one way or the other. As such, this is the greatest criticism by materialists and by people who do not want to associate themself with religion. Here, I would argue that the concept of God isn't necessarily confined to any particular religion. For many, religion is institutions created by people to organize their inner spiritual needs, and for others, it is just one of the ways to come to the idea of God using their culture.

Perhaps, another problem with idealism is that this concept presents a greater challenge for comprehension compared to the notion that life and humanity emerge solely from material processes. The most difficult thing about information existing by itself is how to represent it if there is no direct material carrier. At the same time, the creator and bearer of consciousness are not accessible to our sensitive perception. How can we imagine abstract forms, woven into the very fabric of the Universe?

Intuitive representation is a matter of habit. In quantum mechanics, we have long been accustomed to the idea that an electron or a photon ("quanta of light") behave simultaneously as particles and as waves. Such a habit can easily develop in relation to idealism. Realizing that the primary cause of the Universe's emergence are sense and spiritual beginning, which create the world of things, objects, and human consciousness, is not so utterly unthinkable. At least, compared

---

[3] Transcendent - lying beyond the limits of this word.

87

to the materialistic worldview, according to which the material world just appeared, and then exabytes of information began to emerge literally from nothing for the functioning of complex biological life.

As you have already understood from the content of this book, I am a proponent of idealism or, at least, dualism. In my case, I rely on intuition and understanding of what matter is capable (or rather, incapable) of. I also rely on the incredible precision of abstract mathematical constructs, which describe the physical world with astonishing accuracy. There are numerous examples from history when mathematical abstractions created in the minds turned out to be absolutely correct in describing the world in the future, after many decades and centuries. In the past, the creators of such mathematical constructs used the principle of logic and internal beauty, although no practical application was seen in such reasoning at that time. Therefore, my inclination to consider idealism (or dualism) the correct direction of thought is based on my intuition, experience and knowledge. For others, faith and religion may be required.

It's quite challenging to prove the existence of universal meaningful information that determines the destinies of people, as well as the structure and properties of the material world. However, there's an approach we can borrow from physics. You're probably familiar with magnetic fields. When a pulsation or disturbance forms in a field, we call it a wave. In 1964, theoretical physicists Peter Higgs (1929 – 2024), François Englert, and four other scientists proposed the hypothesis that the entire Universe is filled with an invisible, all-pervading field that imparts mass to elementary particles. The wave or disturbance of this field is called the "Higgs boson". In 2012, researchers at CERN confirmed the existence of this heavy particle. They examined the results of proton collisions accelerated to nearly the speed of light at the Large Hadron Collider (or the LHC). A proton is a stable suba-tomic particle occurring in all atomic nuclei. We will return to this discovery in several places over the course of this book.

This analogy with the all-pervading field is very apt for our further discussion. I will show that in historical records, similar "wave disturbances" can be found in the patterns of information describing past events. They manifest in the peculiarities of some incidents and may resemble "wave ripples" in the biographical facts of individuals. Such phenomena arise at pivotal moments in life and can affect space, time and logical connections between significant events. This is simi-lar to the Higgs boson, which can be observed by colliding two parti-cles at unimaginable energies, leading to rare "wave oscillations" in energy and the possibility of detecting such a particle. In society, sim-ilar effects lead to astonishing repetitions of destinies and coinci-dences that do not obey causal relationships. Perhaps such phenomena are capable of exposing the most fundamental structure of this world, where idealism with its abstract forms is the most plausible explana-tion of the structure of our existence at its deepest level.

3.10 Remarkable number

In this chapter, we will talk about numbers. However, this book is not about numerology — the belief in esoteric or mystical re-lationships between numbers and individuals. Nonetheless, it contains elements that link numbers with phenomena and destinies. There is nothing relating to blind superstition here: We will show exactly where such a connection exists. Although numerology is considered a pseudoscience, this book does not contain any pseudoscientific expla-nations, since it only touches places where there are no any explana-tions at all, which leaves room for fantasies. I explained at the begin-ning of this book that I describe scientific facts and measurements in their current understanding. However, this is not scientific work. This book uses well-established facts, but their interpretation may not be fully rational for materialistic viewpoints. I look beyond the edge of our current understanding, and answer questions for which science remains silent. The role of the scientific method ends at this boundary. Beyond it lie the answers that concern us, the people.

It should be remembered that the founder of the doctrine of the meaning of numbers was Pythagoras (about 570 — 490 BC), an ancient Greek philosopher and mathematician, who is attributed with the statement: *"The world is built on the power of numbers"*. He believed that it is through numbers that the vibrations of the Universe manifest, affecting our entire planet and each individual in particular.

Here I want to talk about the number 6. It is called the "smallest perfect number". In other words, perfect numbers are natural numbers equal to the sum of all its proper divisors. Indeed, the number 6 is equal to the sum of its divisors: 1+2+3 = 6. This concept was introduced by the Pythagoreans in the 6th century BC. According to their numerology, a number's coincidence with the sum of its divisors testified to the special perfection of such a number. Therefore, 6 is the smallest perfect number. The next perfect numbers are 28, then 496, and so on.

In mathematics, the number 6 appears in the most unexpected places. Here is one example: Take any three-digit number. Arrange its digits in ascending order first, then in descending order. Subtract the smaller number from the larger one. Repeat this process with the resulting difference. In no more than 6 steps, we will get the number 495, which will continue to "reproduce" itself indefinitely. Why do these iterations require no more than 6 cycles? Note that for four-digit numbers, our calculation will stop at the number 6174, as demonstrated by the Indian mathematician Dattatreya Kapreka (1905 —

90

1986), although the stopping point will not exceed 7 iterations. For two and five-digit numbers, this property does not exist.

A skeptic might say, "So what? One can always come up with some algorithm that will 'hit' the number 6!". That's not quite true. There are actually quite a few such simple cyclic manipulations with numbers.

If we lived on a line in one dimension ("1D"), there would only be two directions of movement (forward and backward). For two-dimensional ("2D") beings, there are four spatial directions. We live in a world with three dimensions ("3D"), which are measured by length, width and depth. This provides 6 directions of movement. For our world, 6 is a privileged number in terms of spatial movement. That's why the well-known ordinary die has 6 faces with numbered sides. This configuration provides the most balanced behavior for obtaining random numbers and simplicity in manufacturing. Rolling dice is one of the oldest games known to humanity. In ancient times, people believed that the outcome of the random appearance of a certain number was determined by the gods.

The number 6 has immense symbolic value. In biblical numerology (Dare 2017), the number 6 is used to denote man with his fall into sin. According to the book of "Genesis" in the Old Testament, it took God 6 days to create the world. Humanity was created on the 6th day. The sixth commandment says: "Thou shalt not kill". Only 6 generations were given to the offspring of Cain. This number is one less than the number seven, which symbolizes completeness. Therefore, 6 can signify something incomplete or imperfect. This number represents man, with his sin and weakness.

The theologian and philosopher, Saint Augustine (354 – 430), wrote in his famous text "The City of God":

"Six is a number perfect in itself, and not because God created all things in six days; rather, the converse is true. God created all things in six days because the number is perfect.".

However, let's return to modern science. In the Standard Model of particle physics, there are 6 types of quarks and 6 types of leptons. These are the most fundamental and smallest particles that we know of. They have no internal structure. It is from these particles that protons and neutrons are made, which form most of the visible matter.

Many scientists believe that the history of our world began with the Big Bang. Within a short span of time, about 10^{-36} seconds, the Universe was filled with an infinite number of fundamental particles. These were the 6 types of quarks and 6 types of leptons. It was this group of foundational particles that started to interact with each other through other force carriers — bosons. Essentially, this was the "embryo" of our Universe, which began to expand and cool down, forming new particles with masses. These particles are studied at the accelerator at CERN.

The first stars formed hundreds of millions of years after the Big Bang. In stars, the synthesis of heavy elements began as a result of nuclear reactions, starting with carbon. Life on Earth would be impossible without this chemical element. Living organisms, from microbes to humans, are built precisely from carbon compounds. This is related to carbon's ability to easily form long chains and complex bonds with other atoms. This feature gives flexibility to the form of biomolecules from which cells are created. Without this ability, the formation of DNA and RNA, which are necessary for storing and reproducing information in the cell, would be impossible. Carbon is the chemical basis of life, and it is inextricably linked with the very development of human civilization. For example, carbon is a major component of many useful minerals and fuels, including coal, natural gas, and oil.

I would not talk so much about carbon if I had not ended my description of the creation of the Universe with this well-known fact: a neutral carbon atom has 6 electrons. The atomic number of carbon in the Mendeleev table is six. This means that a neutral carbon atom has 6 protons and 6 electrons.

And the number 6 is somehow strangely connected to the sphericity of 2D objects. Six identical circles can be placed around a central circle of the same radius so that each circle touches the central one (and touches both of its neighbors without any gaps). No other number has such properties. Note that the radii of the circles do not matter. Try to perform this experiment with coins of the same size. The densest packing of circles on a plane is related to the number 6.

Solar eclipses, when the Moon completely or partially covers the Sun, have given us a rare opportunity to understand the world. Nature has allotted us roughly 6-minute windows for a total eclipse, during which two great discoveries were made (see chapter 15.4 Solar eclipses). This is the result of the incredible adjustment of the Moon's size, its distance from the Earth and its rotational speed. And since we are talking about measuring time, anyone can notice the division of an hour into 60 minutes and a minute into 60 seconds. This came from the Babylonians, who lived in the early 2nd millennium in the south of Mesopotamia (the territory of modern Iraq). They used a sexagesimal numbering system — a system of numbers based on the whole number 60. Most likely, it was invented by the Sumerians around 3500 BC. The origin of the sexagesimal system is unclear. Perhaps, the Sumerians were so impressed by the number 6 that they started multiplying it by 10 (the number of fingers on the hands).

It must be said that Sumerian culture is the most complex ancient culture in the world. It appeared quite suddenly. According to ancient legends, the Sumerians received knowledge from the gods they called "Anunnaki". They were physically similar to humans but were taller. Possibly, they had six fingers on each hand and foot. Indeed, there are Sumerian engravings and statues depicting their gods with six fingers. The gods taught them everything — even mathematics, which was based on a sexagesimal system. The Sumerians believed that their deities came from the stars — the Pleiades. According to another hypothesis, the "Anunnaki" came from the planet Nibiru, which is in close proximity to Earth (popularized by Von Daniken 1999). Whether this is true or not, the assumption that the number of fingers of the "deities" influenced the use of the number 6 in the Sumerian numeral system is not improbable. After all, the current decimal system originated precisely because humans have 10 fingers on their two hands.

Thanks to Babylonian and Sumerian culture, a 24-hour day is divided into four parts (morning, afternoon, evening and night), each of which has 6 hours. Signs of such a system can even be found on the face of a clock: Just take any number and subtract the number from the opposite side. You will get exactly an absolute value of 6, no matter which number you choose.

A curious reader may ask: What is the likelihood that some number, such as 6, appears in all these places? The Universe was born using fundamental particles in two groups of 6. Later, carbon was formed, containing three groups of 6 particles, which was the building material for life. There is no causal connection between these coincidences. These two events were separated by hundreds of millions of years. Then, the description of the creation of the world in the Bible uses "6 days" for the creation of the world and man. And the Pythagoreans declared that 6 is the "smallest perfect number" due to its mathematical beauty. I think the likelihood of such a coincidence is not very high. These are precisely the questions a person can ask and answer using common sense, experience, and internal intuition. Science will tell you nothing. These questions, by themselves, are deeply philosophical. *"We just conduct a measurement and find a number that needs to be explained. And the same number appears again and again in similar circumstances. Why?"* (Sanger 2024).

In my book, the number 6 will occur exceptionally often. This didn't happen because we selected events where the number 6 is present. When all examples of the most astonishing coincidences and facts were found and described in the draft of this text, this number started to appear quite naturally in the narrative of this book.

4 Do coincidences exist?

In everyday life, we often catch ourselves encountering amazing coincidences. These are remarkable overlaps of events onto each other, perceived as meaningful and related by meaning, yet without an apparent causal connection. Coincidences happen constantly, but people generally do not know how to objectively reason about their probability in everyday life. For example, what is the probability that your birthday is on the same date as your dog's birthday? We will return to the questions regarding the probability of such events later. According to the "law of large numbers", if there are many events and people with whom you interact, there is always a probability that something improbable will happen. This is where most people stop their reasoning.

According to psychiatrist Bernard Beitman, certain personality traits are usually associated with a higher number of coincidences. For example, religious people or those with a high level of meaning-seeking tend to see coincidences in their lives and the lives of those around them. People are also prone to see coincidences when they are depressed or anxious. A likely source of these correlations could be that such people can more easily convince themselves of the significance of the events occurring to them. They may incorrectly estimate the probability of their observations or start to attribute meaning to things that ordinary people do not consider significant. But there is another explanation – such people have a greater connection to the original source of reality. If it exists that is.

In his book "Meaningful Coincidences" (Beitman 2022), Beitman has compiled a vast number of examples of all sorts of coincidences. One of the explanations of the phenomena presented in the book is the connection between people through a Unified Mind. This contemporary study was inspired by the works of the famous Swiss psychiatrist and philosopher Carl Jung (1875 – 1961), who introduced the concept of synchronicity or synchronism.

4.1 Synchronicity

Carl Jung introduced the term synchronicity (or synchronism) in his book "Synchronicity: An Acausal Connecting Principle" (Jung 1973). Synchronicity denotes the coincidence at a specific point in time of several events that do not have common causes. Coincidences in a person's life are absolutely not random, as he claimed. They arise because all meaningful events of nature are organized in some non-physical way. The "higher consciousness" has no spatial and temporal boundaries. Synchronism is just a temporary and often involuntary connection of your local consciousness to the higher mind. Such a connection can occur spontaneously, in moments of grief or pivotal events for a person. And sometimes it is directed by a person practicing spiritual teachings. In the book, he states:

"If space and time are merely properties of moving bodies and created by the intellectual needs of the observer, then their relativization[4] through the content of the psyche no longer represents anything out of the ordinary... The problem of synchronicity has occupied me for a long time, probably since the mid-twenties... I discovered coincidences, so significantly connected that the probability of their randomness was expressed by such an astronomical figure that they were clearly meaningful."

Synchronism suggests the presence of meaning, which is connected with human consciousness and clearly exists outside the person. Such an assumption is contained in the philosophy of Plato, who considered the existence of transcendental forms or models as self-evident. Our world of material objects is just their reflection. According to Jung's theory of synchronicity, certain rare events or coincidences reveal the action of an acausal connection between mental and physical events through their inner meaning.

The essential points of his concept were discussed by Jung with the famous physicist Wolfgang Pauli (1900 – 1958), who received the Nobel Prize for contributions to quantum mechanics in 1945. In their joint book "The Interpretation of Nature and Psyche" (Jung and Pauli 2012), they deepened this principle, allowing us to

---

[4] Making relative rather than absolute.

overcome the gap between mind and matter. Jung and Pauli were convinced that synchronistic events reveal the deep unity of mind and matter, as well as subjective and objective reality.

Psychoanalyst Joseph Cambray further developed this theme in the book "Synchronicity: Nature and Psyche in an Interconnected Universe" (Cambray 2012). He also concluded that synchronicity is a concept that goes beyond the physical world. It touches on the question of the very nature of reality. In situations where synchronicity is observed, the world does not use simple physical laws of cause and effect. Reality is interconnected, just as psyche and matter are. In this world, there are "calls and invitations" in the form of meaningful coincidences that help in life.

We must give credit to another earlier researcher - the Austrian zoologist Paul Kammerer (1880 – 1926) – a documenter of coincidences. Carl Jung based his work on Kammerer's research in his essay "Synchronicity". Kammerer published a never-translated book called "Das Gesetz der Serie" (translated as "The Law of the Series"). He discussed about 100 coincidences, based on personal experience, that led him to formulate the theory of "seriality". He postulated that all events are connected by "waves of seriality". Unknown forces could cause what are perceived as peaks, groupings, and coincidences. Famous theoretical physicist Albert Einstein (1879 – 1955) called the idea of seriality "interesting and by no means absurd".

The meaning of synchronicity, leading to amazing coincidences, is not always clear. The problem is that when such events occur, everything existing in the world participates in them. This affects not only the fates of people but also inanimate objects. Are they signals from some external world?

Arnold Mindell, an American therapist and writer, calls such coincidences "flash-flirts". They are "short-lived, transient, perpetual signals which can be used to provide us with insight". The main idea is borrowed from the life of Australian Aborigines, who believe that surrounding objects can "flirt with consciousness", attracting our attention to indicate something important that needs to be noticed. It's as if the Universe itself is trying to signal upcoming events and inform us about significant changes in life. The mechanism of such flirtations is difficult to explain. Mindell refers to quantum physics for a possible explanation (Mindell 2000).

Deepak Chopra, a well-known American physician and writer of Indian origin, states (Chopra 2005):

"I do not believe in meaningless coincidences. I believe that coincidences are messages, hints that should be given the utmost attention. By paying due attention to coincidences and their meaning, you maintain a connection with the deep layer of infinite possibilities. I call this state SynchroDestiny - it allows you to fulfill any desire. SynchroDestiny suggests access to the deep levels of your essence; moreover, you must carefully monitor the intricate dance of coincidences in the material world. One should try to penetrate the nature of things, and realize the existence of the source of mind, thanks to which the creation of the Universe continues to this day. A person should strive to realize the opportunities opening up before him and thereby change his life. The more attentive you are to coincidences, the more often they happen and the wider your access to messages-hints".

According to his philosophy, coincidences are the key to unraveling the will of the Universe. Coincidences are not a source of meaning. The true source of meaning is the person gaining experience. In coincidences the will of the Universe is manifested, allowing one to use the limitless possibilities of life. By noticing coincidences, you start to ask, "What does all this mean?". By doing this, you can "attract" new information.

Here is one of the statements of those who actively support the existence of synchronicity: At the spiritual level, the past, present and future exist simultaneously. Synchronicity consists of those windows that provide us access to a new spiritual reality beyond the confines of time. Later, we will show that this hypothesis naturally follows from our examination of cases of synchronicity.

4.2 Probabilities of synchronicity

To calculate the probabilities of certain special situations occurring in real life, one must know the number of all possible events (see chapter 3.3 About probabilities). It is not very easy to determine this accurately for events related to people. However, it is still possible to give an approximate estimate for the number of potential possible events.

Almost all of us have encountered incredible coincidences in our lives. They are mainly related to the occurrence of recurring events. Imagine you went outside, and a hailstone fell on your head, even though the sky was completely clear. This event is rare, and its probability is difficult to estimate. All we can say is that such a probability is much less than one. Then imagine that the next day, by some incredible coincidence, a hailstone fell on your head again, although there was absolutely no reason for hail.

The probability that two such events are possible is the multiplication of two small probabilities. Such events are very unlikely, even if we do not know the specific probability for a single (that is, the first in our example) event. Let's consider one example with numbered balls inside of a basket. Let assume you do not know how many balls are in this basket. There could be infinitely many. You started pulling out balls marked with numbers. A blindfold is on your eyes, and you do not know which ball you are pulling out. Having drawn one ball with the number 6, you do not know the probability of such an event. After putting the ball back to the basket, you again pull out a ball with the number 6. In this example, instead of the number 6, any other number could be used. The main point is that it is the same number is drawn in both attempts. We put the ball back. Then we pull out the same number again. The probability that you pulled the same ball on the second attempt is low (although you do not know the exact value, as you do not know how many balls are in the basket). The probability that you will pull out the same ball for the third time is very small. This is because you need to multiply small numbers, being probabilities, to get the overall probability. If you pulled out the same ball again, there could be two scenarios. The first — the number of balls in the basket is very small. The second — you are incredibly lucky (if there are a lot of balls in the basket).

If there are 100 balls in the basket, then the probability that you have drawn any number three times in a row is $(1/100) \times (1/100) = 0.0001$. Note we multiply only two numbers, since we do not care about the number on the first ball. In real life, this is a small probability. However, if you made a million attempts, then the chance that you will pull the ball three times in a row is quite large. This is similar to if millions of people did this experiment at the same time.

In general, our example with the balls is an illustration of the method of calculating probabilities for events in society. This method, where we ignore the probability of the very first event, since often we do know much about its likelihood, gives us the most conservative estimates for the values of event probabilities. The main point of this approach is that there are several recurring cases, and they are unlikely to happen. It means we simply multiply many small numbers (probabilities) to find the full probability of all such events (see chapter 3.3 About probabilities).

Let's illustrate this method with a real example taken from life. Anne Hathaway is a famous American actress and singer. The wife of the English poet and playwright William Shakespeare (1564 – 1616) was also called Anne Hathaway (1556 – 1623), that is, the name and surname exactly coincide. This is an interesting coincidence, and perhaps even the probability of such a coincidence can be calculated. However, instead, we will use such a case as a "starting point", which drew our attention to this coincidence. Often, we do not know the number of possibilities for the first coincidence. All we know is that it attracted media attention. Therefore, many people started discussing it. On an emotional level, such a coincidence seems amazing and almost impossible. Therefore, many people begin to look for other coincidences. And they find them. In this example, such a coincidence was indeed found. It turns out that the face of Anne Hathaway's husband looks very similar to the face of the playwright William Shakespeare (at least, compared to the portrait of William Shakespeare made after his death). According to some research (Gorvett and Brunelle 2016), the chance to find a pair of men with similar faces is approximately one in 50,000. Actually, I think this probability is somewhat larger, as it is difficult to identify some features of William Shakespeare's face from the portrait if this picture is indeed accurate in describing the playwright. In this case, I reduced 50,000 to 5,000.

Thus, the probability of the coincidence where two women with the same surname are married to men who look similar is 1 / 5,000 = 0.0002. As we can see, we completely ignored the probability that the two women have the same surnames and names.

There is something else unusual in this example. Both families, where the women are called Hathaway, are associated with theater and acting. Shakespeare's wife could not have been an actress, as women actresses were rare at that time. However, there is a meaningful connection to acting through her husband. I cannot judge the exact probability of the "encounter" for two families that are professionally connected. Based on my experience, the probability of 1/10 looks realistic. Therefore, the probability that there are two women with the same names, whose husbands look similar and are related to acting, is $0.0002 \times 0.1 = 2 \times 10^{-5}$. This value does not consider the probability of the coincidence of the names of these women. This makes sense, as this initial moment of coincidence was simply a "trigger" for our attention to this case.

As we will show, emotional expectations (or other feelings) can indeed cause a series of other coincidences. In this example, the faces of the two men turned out to be very similar. We are specifically interested in these secondary coincidences and their probabilities. We will show that secondary coincidences, in some situations, are not random. Most likely, they are caused by people's expectations and their emotions. These anticipations lead to the effect of synchronicity. How exactly this effect occurs, we do not know. One of the likely hypotheses relates to the editing of information about the past that has reached the observer in the present time. This is how such effects, but related to microscopic particles, can be interpreted in quantum physics. We will discuss this topic in chapter 8.2 Nonlocality.

The example discussed above was not associated with strong tragic experiences. Therefore, there are some doubts about whether synchronicity occurred. I will provide another example, taken from the book of Michael Shermer, an American science popularizer and founder of the "Skeptics Society". I found his story quite interesting and trustworthy, as it was told by a person focused on investigating pseudoscientific and supernatural claims. He describes this example of a strange coincidence in his book "Heaven on Earth" (Shermer 2018). The narrative goes like this: His wife was in a state of depression, being away from home and reminiscing about her deceased grandfather. Suddenly, she heard a nostalgic love melody from the old radio sitting on the shelf in the bedroom. This radio was non-operational, but it was dear to the woman because it reminded her of the grandfather with whom she used to listen to radio broadcasts. In the past, Shermer tried to fix the receiver but failed. So, he just left it in the turned-on state on the shelf. How did the receiver start working by itself at the moment when moral support was needed, and it did not just start working — it began playing a melancholic melody?

This case indeed looks like a typical synchronicity, as pointed out by American physician and writer Deepak Chopra. However, a quantitative estimation of the probability of such an amazing event has not been made. Let's try to make an assessment of this coincidence. First, we do not know the probability that a non-working radio will spontaneously start playing music. But this is not important to us, as it is just a starting point for the calculation. Suppose the radio turned on. The probability that it tuned by itself into a frequency where a certain radio station operates is approximately 1/10. I just checked this for the Chicago suburbs, where stations quite densely fill the frequencies. Further, the probability of whether they broadcast music or not is 1/2. Whether the music is sad or cheerful — 1/2. The reason for 50% chance is because we do not know anything about the correct likelihoods for such circumstances. Therefore, the final probability value can be estimated as: $1/10 \times 1/2 \times 1/2 = 0.025$ or 2.5%. This is not a very small probability, but nonetheless, this case deserves attention. This probability corresponds to the situation when the radio randomly was set to a certain wave frequency with a sad melody, assuming that the radio randomly turned on before that happened.

The main problem with this example is that this event did not occur at a very rare and pivotal moment in life. Many people experience sadness very often, so the number of possible situations (or possibilities) can be quite large.

I will provide another case sent to me by my friend Tim. He decided to visit his neighbor. The last time he had spoken to her was a month ago. He approached her door and found out that she had just been typing a text message to him. She did not know that Tim was coming to visit her. Suppose "just now" refers to a 10-minute interval from the moment Tim rang her doorbell. The number of such intervals in a month (consisting of 30 days) is exactly $30 \times 6 \times 24 = 4320$. Therefore, the probability of a random coincidence is $1/4320$. This is a small probability. The difficulty in calculation here is to understand how many such situations occurred in Tim's life, that is, the total number of possibilities. I note that cases where you call someone, and the person responds that they were just thinking about you are exceptionally frequent. I myself have witnessed such situations multiple times.

Using these three examples of synchronicity for any significant conclusions is difficult. There are approximately eight billion people living on Earth. Indeed, rare coincidences are possible for so many people. In the following chapters, I will show how to reduce the "large numbers" problem when calculating synchronicities.

4.3 The birthday problem

Now let's delve into the intriguing paradox known as the "birthday problem", which may often cause confusion among people. It is quite appropriate here to provide the reader with an intuition about how frequently various coincidences can occur.

To illustrate that coincidences are fairly common when there is a sufficiently large number of people, let's answer this question: "How many people need to be surveyed on average to find two people with the same birthday?". Apparently, such coincidences can occur with a higher probability than we might think. It appears that if you consider a random group of just 23 people, the probability that two of them will have a birthday on the same day is about 50%. Just 23 people! This is known as the "birthday paradox".

These calculations can be done analytically. But since we have decided to use programming code, let's try to use computer calculations. We will assume that birthdays can be repeated. Using the programming code in 20.4 Appendix, we get the answer:

- Probability of > 0 pair = 0.506
- Probability of > 1 pair = 0.144
- Probability of > 2 pair = 0.032
- Probability of >3 pair = 0.0072

Thus, the probability that at least 2 people have a matching birthday is approximately 51% (with repeating birthdays). However, the probability value drops to 0.72% if more than 3 matching pairs are required. Increasing 23 to 100, we will see that the probability of a match for two birthdays will be close to 100%.

In our family, including grandparents, there are probably 23 people. But we do not have a single match for our birthdays. Perhaps, we fall into those 50% where this coincidence simply does not occur. Nevertheless, we have two coincidences when one birthday matches a day of death. My father died on my mother's birthday, and the daughter of my mother's brother was born on my grandmother's birthday. Thus, we have two coincidences where birthdays correspond to days of death. We can calculate the probability of such a coincidence. This can be done quite simply using the previous program code:

- Probability of > 0 pair = 0.95
- Probability of > 1 pair = 0.77
- Probability of > 2 pair = 0.51
- Probability of > 3 pair = 0.21

Based on this, coincidences between birthdays (or deaths) are quite common events for a group of 23 people. These probabilities are quite large since we increased the number of possibilities.

All these examples suggest that even if you have noticed some coincidences between birthdays or days of death, this says very little about anything unusual. You should, at least, estimate the number of people involved in your experiment. If this number is more than 20, then the likelihood of coincidences is very high.

Another problem is that it's not always clear what the significance of such coincidences is. The times of birth and death are indeed pivotal moments in people's lives, but the logical context is not very clear. First, they are not tied to any significant (other) events. As we will show below, being tied to significant events is a condition for the occurrence of rare events. Secondly, distant relatives are not sufficiently unified in meaning (only by biological kinship). If the relatives worked in the same field and were the closest (like a daughter and mother, or husband and wife), then this could create the necessary effect, which would increase the possibility of unlikely coincidences.

4.4 Our close-knit world

On the internet, you can find quite a lot of stories describing situations with incredible coincidences. In most cases, people try not to disclose their names. Here I will tell one story with an amazing coincidence (Dobrogosz2024) for which one can estimate the probability:

"When my nephew was little, he always played with the home telephone. One day he took the phone from my grandmother and started pressing some numbers. Then he started talking to someone. Seeing

this, my grandmother took the phone to apologize to whoever it was. It turned out to be my grandmother's cousin, who lived on the Isle of Wight, with whom she had lost touch many years ago. My nephew accidentally dialed his number!"

Such a coincidence indeed looks very unusual. Let's calculate the probability that this event happened by chance. In the USA, phone numbers consist of 10 digits. Therefore, the total number of possible combinations is 10^{10}. The probability that a child dials some ten-digit number and calls a person known to the grandmother is 10^{-10}. Is such an event surprising and worthy of our attention?

The population of Earth will soon reach 8 billion. But we are interested in North America since this news appeared here, not in other countries where I do not read the news. Also, we are interested in the approximately 20 million children of North America who could potentially play with the phone, not realizing who they are calling to. Then it should be considered that this grandmother might have several relatives. Let's assume that there are 10 relatives. If each of the 20 million children called some phone number, then the probability that someone will get through to 10 known numbers is:

$$10^{-10} \times 20 \times 10^6 \times 10 = 0.02.$$

This is not such a small probability. And if we consider that each of the children randomly dials several numbers, then we are talking about a value of about 10–20% for the probability that such a case occurred at some point of time.

In this example, we again deal with a very large initial sample of people. 20 million is a sufficiently large number, which greatly increases the number of possibilities. And what if the selection was not random? Perhaps the grandmother simply forgot that she has this relative, and she had communicated with them before? The child could just dial the number by the remembered tone of the buttons if the grandmother had already called this person (but she forgot this). But most importantly — we are dealing with an anonymous source, so it's quite difficult to verify the legitimacy of this case.

5 How to crack the code

Probably, many of us have encountered situations where events that happened to us, or our acquaintances seemed very improbable. Such incidents can appear so unreal that we couldn't help feeling that this world is not what it may appear on the surface. It is not empty and cold, where only blind chance and physical laws rule. In the previous chapter, I provided several examples of incredible coincidences. You yourself have likely also heard many stories about all sorts of unusual coincidences and omens. Although we wanted to believe in such stories, somewhere deep in our consciousness, many of us didn't fully trust these narratives because rational explanations did not exist.

Maybe, this world and all of us are influenced by an intelligent force that violates all accepted laws of causality. But how can we prove it?

The main reason why many people do not trust these amazing stories about coincidences lies in a simple explanation: Among the vast number of people living in this world, something unusual can always happen at some time and in some place. This is one rational way to think about this: Using the terminology of probability theory, we do not know the total number of potential or possible events. See chapter 3.3 About probabilities for more. In addition to this, we often learn about rare events involving people after they have already happened.

However, there is a way out – instead of considering all possible events with all the people with whom these events happened, we can focus on a deliberately selected group of people.

5.1 Probabilities and possibilities

Let's recall the example with balls considered in chapter 4.2 Probabilities of synchronicity. Suppose there are 100 balls in a basket. The balls are numbered from 1 to 100. If we randomly pull the same ball 6 times in a row, having previously put this ball back before each drawing and blindly mixing the balls, then the probability of such an event is $(1/100)^5$ or 10^{-10}. Note that we don't know what the first ball's number is. If you observe one randomly selected person who takes balls out of the basket with their eyes closed, then drawing the same ball 6 times in a row is an exceptionally rare event. The numerical value of the probability of such happening is very small.

On Earth, there are about 8 billion people (as of 2024), so rare coincidences are possible due to the principle of large numbers. For example, pulling out the same ball six times out of a basket in sequential order is not a rare event in the context of such a large number of people. The statistical sample (that is, the number of people with whom something incredible could have happened) is very large. The probability to observe such an event for 8×10^9 people is:

$$8 \times 10^9 \times 10^{-10} = 0.8.$$

As we see, this number approaches 1. This means that someone is bound to pull the same ball 6 times in a row. We will surely hear about this news from the press and marvel at such luck.

Let's give another example. If a huge number of coins are tossed onto the floor, there is a real probability that one of the coins will land on its edge and remain in that position. This probability has even been calculated in the past (Murray and Teare 1993). There is a 1 in 6000 chance that a coin will land on its edge.

However, imagine a situation where we do not know how all the coins land on the floor. But it is within our power to check how 100 coins, marked by us before they were dropped, will land. And it turns out that one of the marked coins actually landed on its edge. Of course, such an observation would be quite incredible. It leads to the assumption that very rare events happen much more frequently compared to our expectations, based on empirical-rational perceptions of the world. But if you throw the coin again, and it lands on its edge again, then there is no doubt left — there is something missing in our understanding.

Let's imagine such a situation again. All the people on Earth decided to toss a coin twice at random moments of their lives. Among 8 million people, there will be those whose coin landed on its edge twice. The probability of such a coincidence, as we know, is very low — $(1/6000)\times(1/6000) = 256\times10^{-10}$. Newspapers will write about these people, so we will learn about them. The number of such people can be calculated — it will be about 205 people (multiplying 256×10^{-10} by 8 billion).

Now imagine a small village, say, of 100 people. This village will be our control sample. We are not interested in anyone on Earth except the people from this village. These people also toss a coin twice in a row. But we, the observers, can only see these tosses at a specific moment — only when they get married. By choosing this village, you start to observe people with the aim of seeing the results of the coin toss. And it turns out that there is a person from this village whose coin landed on its edge twice during the time period we observed. In this case, it can be said without exaggeration that this is a very strange village since the occurred event is improbable. This incident cannot fit into the framework of statistics from the point of view of an observer who decided to oversee this village for reasons not related to coin tossing. But even more astonishing is that this happened within a short period of time, which we determined in advance — after a specific event (marriage).

To understand how strange this looks, take the rest of the Earth's population and create 80 million groups of 100 people each, and track how they tossed their coins. Of course, you will find a group where the coin landed on its edge twice after you surveyed everyone. For you, this will be some random group, since you did not "mark" it in advance using any specific criteria. Therefore, you will not see anything unusual in this, as you already know how many groups are participating in the experiment. But if you randomly decide to find a group where a coin landed on its edge twice, then, most likely, you will not succeed. It is even less likely to find groups where the event happens at a specific time.

Another problem will be as follows: When surveying all the groups, you will surely doubt the truthfulness of the stories from people whose coin landed on its edge twice. Unfortunately, there was no objective and independent control over the coin tossing.

5.2 Significant events

So, we will only be interested in unusual incidents that occurred with a limited number of pre-selected people. However, we will require that such events should be their most important life events, or "turning points" in their fates. For example, such events can be birth, death (if it occurred naturally), marriages and various accidents after which people's life changes dramatically.

Why is it necessary for such unusual events to be the most significant in the lives of people? First, there is a statistical reason, being that it is done in order to limit the number of potential situations. Secondly, it is anticipated that intense emotional states can create conditions that fully facilitate the manifestation of synchronicity phenomena. In the article "Synchronicity and Healing" (Beitman, Celebi, and Elif Coleman, 2010), it was found that experiences of synchronicity mainly cluster around periods of emotional stress or major life transitions, such as birth, death and marriage. Perhaps it should be noted that there are other earlier experiments that showed unusual effects at moments of emotional stress. We will assume that if publicly documented refutations do not exist, then such experiments deserve our attention.

One such experiment was conducted quite some time ago. In 1986 - 1995, French researcher René Peoc'h (Peoc'h 1995) conducted the following experiments: He created a robot in the form of a small tin can with wheels, which received random direction impulses as it moved. It had a random number generator at its top to achieve this task. Every few seconds it changed direction. As it moved, the robot made random turns so that its trajectory was completely unpredictable. The robot had an attached pencil used to trace its path on paper. A large table with a low side wall, preventing the robot from falling off the edge, was covered with a sheet of white paper. The robot was placed in the center of the table and set in motion. As a result, it began to draw a random pattern on the table.

Then Peoc'h released newly hatched chicks onto the platform with the robot. Chicks, like all birds, believe that the first living creature or object they see after hatching is their mother. Therefore, the chicks started to follow the robot around in a random trajectory.

Next, the experimenter placed the chicks in a cage at one end of the table, placed the robot in the center of the table, and turned it on. The chicks began to feel anxious when the robot moved away from them. It was assumed that the chicks would want the robot to be as close to them as possible. The robot again began to draw a random trajectory on the table, which was significantly closer to the cage with the chicks.

When Peoc'h put rabbits in the cage, they were scared of the robot, and the robot started moving to the far end of the table. But when the rabbits got more comfortable with the robot, it started to approach them. The scientist rechecked the experiment many times, but the results were the same.

The reason why I decided to talk about this experiment is not only because it is interesting in itself. The point is that currently, there are no independent checks confirming (or refuting) its results. Given the simplicity of such research, this seems strange. But if this experiment was not erroneous, then it fits perfectly with the following hypothesis: emotional moments can influence matter and meaning, leading to events that cannot be described materialistically.

As we have already said, we have not found independent confirmation of the results of the previous experiment. Therefore, its conclusion can be taken as a working hypothesis. As we will show later, even incredible fantasies can easily become reality.

5.3 The number possibilities for synchronicity

So, we've hypothesized that strong emotional experiences increase the likelihood of unexpected, statistically improbable events. For the average person, there are unlikely to be more than 10 significant or pivotal events upon which life or one's future fate depends. These could be events associated with significant emotional experiences. For example, birthdays, deaths, weddings, the birth of children, graduation from university, getting one's first job, and so on. All dates associated with such events are perfect for our consideration. This approach to selecting important events significantly narrows the number of possible events for calculating the probabilities of synchronicity.

For each significant event, there are about 30 possible connections to other important events. They correspond to coincidences of a given important event with some other potential and significant incidents in one's destiny. For example, for a wedding day, possible options include:

- A coincidence of the wedding date with significant dates in the future or past. For example, with the birthdate of a child born from this marriage.

- An unexpected meeting with a person with whom there is a special connection.
- A coincidence of the surname with one of the guests, who is of significance for the future family.
- A catastrophic event that somehow is meaningfully connected to the wedding day,

and so on. Why 30? Of course, this is a guess. I simply took a piece of paper and wrote down possible coincidences between a given specific event and other important events. My imagination ran out after number 22. Therefore, the number 30 is a good conservative estimate for synchronous connections. In fact, there might be far more connections than significant events themselves (10 for our calculations). Of course, such connections must exist also with significant events, names, dates and other things. In our example with the wedding, a coincidence of the restaurant manager's surname with the bride's name does not constitute an important connection. Nothing depends on such coincidence for this bride, unless this restaurant manager starts playing a significant role in her destiny in future. Therefore, this coincidence cannot be included in these 30 event connections.

So, we have 10 life-changing events, and for each, there could be approximately 30 synchronous, connected events. Then, the total number of possibilities for synchronicity events during a person's life is:

$$10 \times 30 = 300.$$

This is an important number which will be frequently used. We will call it "the number of potential possibilities for synchronicity throughout the life of a single person".

To understand whether synchronicity occurred in a person's life, we must multiply the probability of a specific case of synchronicity by this number. In chapter 4.2 Probabilities of synchronicity, we looked at two examples of possible synchronicity. In the case of Anne Hathaway, we got a number of about 2×10^{-5}, which corresponds to the probability that two women with the same names have husbands with similar faces, and that their families are connected by one profession. Thus, the probability that coincidence played a role in the synchronicity of Anne Hathaway is equal to:

$$300 \times 2 \times 10^{-5} = 0.006.$$

This indicates that Anne Hathaway's life is likely subjected to synchronicity since the chance of a random coincidence is only 0.6%. This value is slightly less than the probability of getting "heads" (or "tails") 7 times in a row when tossing a coin.

The case with the radio receiver, described by Michael Shermer in chapter 4.2 Probabilities of synchronicity, could be explained by chance. Indeed, $300 \times 0.025 = 7.5$. This number is much greater than one. Even if the described moment of sadness is included in the number of significant events, this event is not special, considering the existence of 300 possibilities.

In general, the number 300 can easily be used for a simple evaluation of synchronicity. If your case of coincidence without an apparent causal connection can be assessed as one chance out of 300 (or less), then this event is not statistically surprising.

5.4 Control sample effect

For our experiment, we need an "observer" who follows a certain number of people. These individuals should be selected based on some principle that is not related to the experiment that we are conducting. We will call this group of people a "control sample" or "control group". It's feasible to limit ourselves to a small sample of individuals who are well-known and whose biographies are reliably documented.

If you can name 100 famous people, it means your memory is in good shape. We need roughly 100 people, chosen in advance not because something amazing happened to them, but because they ended up in your memory for some other, unifying reason. For a group of people of this size, flipping a coin on its edge is a rare event, with a probability value being close to zero. If such a thing really happened — it would be an astonishing event for such a small group of people.

This is the essence of our approach. We should examine the biographies of the most famous people. Their life stories are usually well described by independent biographers. This will be our control sample. Then, we will look at the pivotal or life-changing events in their biographies. Very often, such events are significant not only for them but for all of society as well. Let's give some examples.

Example 1. Imagine you know a famous person because of some, unrelated to synchronicity, reason. You find in their biography, which is reliably documented by third parties, a situation like this: The person's spouse dies on May 2nd of a certain year. Then we learn that 2 of their children also die on May 2nd, but in different years and for completely different reasons. Conspiracy or any foul play is ruled out. We calculate the probability of this happening as $(1/365) \times (1/365) = 7.5 \times 10^{-6}$, where 365 is the number of days in a non-leap year. This is a very small probability, indicating it cannot lead to the observed situations assuming that everything in the world happens by chance. Also, note that these events are pivotal and definitely fall into the 10 most significant events for this individual.

Example 2. A famous person with a well-documented biography has a son. They named him Roman. We assume that this name is among the 20 most popular names. Another well-known person also has a child, and they too name him Roman. Both famous individuals are connected by some thematic category. For example, they are both biologists. Their children are born in the same year and the same month, say, in April. At the time of the birth of both children, it suddenly starts to snow, which must be a rare occurrence for April. Let's assume that the probability of snow in April is well-known — 0.0001. What are the chances that such an event is possible? We calculate: $(1/20) \times 0.0001 = 5 \times 10^{-6}$. This is a small probability, considering that both families belong to the group of 100 people we are observing.

To determine whether such events are synchronicities or lucky coincidences, one must multiply the probabilities calculated earlier by 300, as we discussed in the previous chapter. If the resulting numbers are much less than 1, then synchronicity was likely present. In these two hypothetical examples shown above, synchronicities were indeed present. This assertion will only be valid if these individuals belong to our control group, predefined in advance (and not at the moment when we learned about these amazing coincidences).

Let's summarize. Spectacular coincidences are always possible if one learns about them after they have occurred. This is because there are a vast number of possibilities related to the fact that we are dealing with too many people. But if one focuses attention on a small group of people, selected based on some criterion not related to synchronicity, then this changes the situation completely: Such phenomena become significant for our conclusions.

For example, if incredible events began to happen to all your childhood friends at exceptionally important moments of their lives, this could be interpreted as a certain signal. It is not a coincidence —

it is a pattern that needs to be studied. This is what we will do. In the next chapter, we will look at a group of famous people whose fame came to them long before oddities began to occur with them at important moments of their lives.

Incredible coincidences are sometimes interpreted as omens. Such phenomena convey information to a person or group but are only understandable within the context of specific events. We will return to this in the following chapters.

6 Examples of significant coincidences

In this chapter, we will examine several well-known coincidences and assess their probabilities of occurring due to sheer chance. We will strictly follow our approach outlined before — unusual events must happen to well-known individuals. And the events themselves must be pivotal in the lives of these people. We will not provide a list of the top 100 famous people, as it could be contested. Nevertheless, most individuals discussed in this chapter should undoubtedly be included in this list.

6.1 Abraham Lincoln and John Kennedy

We will examine the circumstances of life and death of the American statesmen and political figures Abraham Lincoln (1809 – 1865) and John F. Kennedy (1917 – 1963). Both politicians are unquestionably among the hundred most famous people. Moreover, their lives are well documented. Also, both were politicians, indicating a certain logical connection between them.

Lincoln became president in 1860, and exactly one hundred years later, Kennedy was elected president. Although 100 is an interesting number, we cannot use it as we will not engage in numerology (i.e. a belief in mystical numbers). Therefore, we will consider some other coincidences. In the calculations, we will use the method described in chapter 3.3 About probabilities.

1. Both presidents were assassinated on a Friday. Public events are usually held on weekends or on the eve, that is, on Friday. The probability of such a coincidence is 1/3, where 3 corresponds to the three possible days of the week (Friday, Saturday and Sunday).

2. The spouses of both presidents lost a child while living in the White House. We do not believe that this fact has a small probability. Without reliable information on how possible events are distributed, I will assume that their probability is the same. In this case, we have only two choices - a child dies during the presidency or not. Therefore, the probability is 1/2 = 0.5.

3. Both assassins were Southerners with extremist views. Here too, we have no preference for whether they are Southerners, Republicans, or Democrats. Therefore, the probability will be 0.5 (Southerner or Northerner).

4. Both assassins were themselves killed before trial – 0.5 (before trial or after).

5. The successors of both were Vice Presidents named Johnson, who were Democrats and former Senators. In America, the 5 most popular surnames are Smith, Johnson, Williams, Jones and Brown. Johnson is one of the popular surnames. Ignoring all other possibilities, our probability is 1/5 or 0.2.

Now let's calculate the total probability of these five coincidences:

$$(1/3) \times 0.5 \times 0.5 \times 0.5 \times 0.2 = 0.0083$$

or about 0.83%. In this and future instances, we will round numbers to show their approximate values. As we see, the probability that such events could occur with one pair of people is quite small.

Note that there are about 16 (some even see 20) coincidences in the fates of Lincoln and Kennedy, but it's difficult for us to estimate their probability. However, there are three more coincidences that truly boggle the mind. Look at the number of letters in the names of people involved in those events:

- LINCOLN and KENNEDY — seven letters (7).
- ANDREW JOHNSON and LYNDON JOHNSON — thirteen letters (13).
- JOHN WILKES BOOTH and LEE HARVEY OSWALD — fifteen letters (15).

The probability of such a coincidence is expected to be small. Suppose we have a lottery with two baskets, where there are balls numbered 1 to 15 (the maximum number of letters). Drawing a ball with the number 7, we may ask: What is the probability that the next ball from the other basket also has the number 7? This leads to a probability of 1/15 = 0.0666. We put both balls back, and then we draw a ball with the number 13 from the first basket. Surprisingly, we again draw a ball with 13, but from a different basket. And then we do the same with the ball numbered 15. Let's calculate the probability of these three matches:

$$0.0666 \times 0.0666 \times 0.0666 = 0.0003.$$

This isn't quite an accurate assessment of probability. In fact, it does not take into account other situations, such as matches by surname. For example, the number of letters in BOOTH and OSWALD do not match. This means that the number of all possible events that might match is much greater than we assume. However, the surname JOHNSON matched exactly. Even considering that JOHNSON was a popular surname at the time, this may compensate for some overestimated likelihood for finding the correct number of letters.

Now let's calculate the total probability of the coincidence that all 6 matches occurred (including matches in names):

$$0.0083 \times 0.0003 = 2.5 \times 10^{-6}.$$

This is a very small probability for a single life-changing event, consisting of the totality of all the coincidences I described. However, the obtained probability does not take into account the many possibilities for coincidences in other important events of their lives. Therefore, we will multiply this probability by the number of possibilities for synchronicity (300) from chapter 5.3 The number possibilities for synchronicity.

$$300 \times 2.5 \times 10^{-6} = \mathbf{7.5 \times 10^{-4}}$$

We highlighted the final result in bold. This probability is approximately equal to 10 consecutive occurrences of "heads" (or "tails") when flipping a coin. It provides an estimate of the probability that

128

blind chance could create synchronicity in the life of Lincoln (or Kennedy) in the form of a group of several coincidences. The smallness of this value may indicate a certain mechanism in the appearance of all these coincidences simultaneously. In principle, such similar circumstances are difficult to observe in a sample of several hundred well-documented biographies of famous people.

This is one of the interesting cases of synchronicity. It manifested several times, like waves radiating from the main source — people's expectation that the fates of Lincoln and Kennedy should be somehow similar. What caused such expectations, the political views of both, 100 years separating their presidency, or a few other initial random coincidences, is not very clear. As this is a fairly well-known case, we will discuss its possible mechanism in chapter 16.4 Symbols and forms of ideas.

6.2 Stephen Hawking

Theoretical physicist, cosmologist, and writer Stephen Hawking (1942 – 2018) rightfully takes his place among famous individuals. Thus, he can be used as a "marker" in our approach to calculate the probabilities of improbable coincidences for significant events. Dates of birth and death are the most significant numbers for people's fates. Hawking was born on January 8, 1942, exactly 300 years after the day of death of the famous Galileo Galilei (January 8, 1642). Hawking died on March 14, 2018, which would have been exactly 139 years since Albert Einstein was born (March 14, 1879). Here, there are 2 coincidences:

- Hawking (died on March 14, 2018) —
  Einstein (born on March 14, 1879)

- Hawking (born on January 8, 1942) —
  Galileo (died on January 8, 1642)

Hawking, Galileo and Einstein are a group that falls into the category of famous people linked by their occupation (physics). If the coincidence had been between the dates of birth or death of Hawking and, for example, a famous biologist, then such a coincidence would have been less interesting. I have nothing against biologists, but individuals who are not directly related to Hawking and have not had a professional influence on his work cannot participate in our calculation. Birth and death days are also significant dates. If Hawking's death day had coincided with the day Einstein changed his residence, then this coincidence would not have sparked any interest for anyone. We are looking for equivalently important dates.

Considering that the three physicists are among the top ten famous scientists, and we are dealing with two significant numbers (birth and death), we can estimate the probability of the coincidence that two significant dates of one person matched with equally significant dates of physicists who were related in profession.

This example reminds us of our consideration in chapter 4.3 The birthday problem. However, we have two coincidences, not one. In chapter 20.5 Appendix, we calculated the probability of these two coincidences for three people. The probability that the birthday (or death day) of one of them coincides with the other two is:

$$6.234 \times 10^{-5}.$$

This is the probability of a double coincidence due to statistical randomness. Now, we find the probability that some synchronicity occurred with Hawking by chance. We will multiply this probability by the number of possibilities for synchronicity events (300) from chapter 5.3 The number possibilities for synchronicity:

$$300 \times 6.234 \times 10^{-5} = \mathbf{0.0187}.$$

The obtained probability, highlighted in bold, is approximately 1.9%. This is a relatively small chance for a random coincidence. It is equivalent to flipping a coin 6 times in a row and having it land on heads (or tails) each time.

In this example, we ignored the fact that Hawking was born exactly 300 years after Galileo's death. For us, 300 and 139 are values of equal significance. We will not engage in numerology in this chapter.

This example of synchronicity is not particularly impressive. If there was synchronicity in this case, it would have been quite weak. Certainly, physicists do not influence people's emotions and well-being to the same extent as politicians who initiate wars. And as a consequence, the emotional effect of scientists' deaths on people was not as strong as in the case of politicians. It's worth noting that synchronicity events can manifest outside time. Their "waves", leading to repetitions and logical connections between events, may go into the past and present. I cannot say what the primary coincidence in this example was that led to the secondary coincidences and the adjustment of parameters in birth dates.

6.3 Richard Bach: The Story with the Biplane

Richard Bach is the author of bestsellers like "Jonathan Livingston Seagull" and "Illusions: The Adventures of a Reluctant Messiah". He is renowned for his love of flying small private airplanes, which is why almost all of Richard Bach's works touch on the theme of flying in one way or another.

In his book "Nothing by Chance", written in the 70s, he described an accident with his airplane. At that time, Richard Bach was flying a rare biplane, the Detroit-Parks P-2A Speedster, one of only eight built. This plane was designed before any aviation standardization. While flying over the Midwest, the plane broke down. After landing and inspecting, the repair seemed hopeless due to the rarity of a needed part. As Bach pondered over this problem, a stranger who owned a nearby hangar approached and asked if he could help. Bach told him about the broken part in his airplane. The man replied that there might be some parts in his hangar. He went over to a pile of metal parts in the corner of the hangar and immediately found the necessary part!

What are the chances of such a coincidence? In the 1970s, there were about 100,000 farmers in Wisconsin, and surely each one had a barn with some mechanical parts. Assuming that all parts from the 8 existing rare biplanes were only in Wisconsin, the probability of the event that the plane would land near a barn with the needed part is calculated as $(8/100,000) = 8 \times 10^{-5}$. We assumed that the likelihood of locating some mechanical part is uniform across all barns.

Here, I note that I wanted to check how unique the airplane part found in the Wisconsin barn was. Could it be that some airplane parts are quite similar to each other? As it happens, Bach resides quite close to me; However, due to his advanced age, he does not grant interviews and was unable to respond directly to my inquiries. Saying this, we should keep in mind that I assumed that all biplane parts should be in Wisconsin, but not in other US states. This compensates for the effect of possible non-uniqueness of the airplane part in our calculation.

So, was there synchronicity in Bach's life? Let's multiply the obtained probability by the number of opportunities for synchronicity (300) from chapter 5.3 The number possibilities for synchronicity:

$$300 \times 8 \times 10^{-5} = \mathbf{0.024}.$$

So, blind chance could have created synchronicity in Bach's life with a probability of 2.4%. Of course, it could be argued that we have incorrectly estimated the probability of such an event with the airplane breaking down. However, here I will rely on the opinion of Richard Bach himself. He wrote about this in his book:

"The odds against our breaking the biplane in a little town that happened to be the home to a man with the forty year old part to repair it; the odds that he would be on the scene when the event happened; the odds we'd push the plane right next to his hangar, within 10 feet of the part we needed - the odds were so high that coincidence was a foolish answer".

I mentioned Richard Bach being a philosopher for a reason. Here are some of his words:

"One of the great cosmic laws, I think, is that whatever we hold in our thoughts will come true in our experience. When we hold something, anything, in our thoughts, then somehow coincidence leads us in the direction that we've been wishing to lead ourselves".

This too is one of the cases where the manifestation of synchronicity is not very strong, as only a few people were involved. What attracted me was the fact that Bach understood the essence of the philosophical problem and recorded it in his book. Perhaps this triggered the synchronicity and the editing of the past — namely, the description in the text of his book, which states that the airplane part was very rare. When he wrote the book, this incident struck everyone involved in those events with its rarity, that is, the plane landing right next to a barn where there happened to be a specific airplane part. Of course, not every barn has airplane parts. This became the focal point of the beginning of synchronicity. Perhaps the initial emotions and astonishment led to a slight change in the story of the past. Namely, the memories of all who participated were "edited", that is, new information emerged about the extraordinary rarity of this airplane part. This enhanced the effect of the unusual coincidence.

6.4 Gravediggers of the USSR

In 2022, when the USSR was commemorated 100 years after its formation, several politicians who took direct part in the signing of the official documents that led to the dissolution of the USSR, passed away. By this time, Boris Yeltsin, who was a key figure in the event of the USSR's dissolution, had already died. However, Yeltsin's closest associate and the main ideologue behind the collapse of the USSR, Gennady Burbulis, died in June 2022 at the age of 76. Burbulis participated in the signing of the Belavezha Accords, which dissolved the USSR. Other participants in the signing of the dissolution of the USSR, the former head of Belarus, Stanislav Shushkevich, and the former President of Ukraine, Leonid Kravchuk, also passed away, but slightly earlier — in May 2022. At the time of death, both were 88 years old. Then, on August 30, 2022, Mikhail Gorbachev, the last leader of the USSR, who was the focal point of the last days of the USSR, died at the age of 92. He was against the dissolution of the USSR, but it was his political weakness that allowed this event to happen.

This case also falls under the conditions of our principle, where we select the most famous people and their most important events. Gorbachev is a significant figure, and the dates of death are also important numbers. All four politicians were directly (or indirectly, like in the case of Gorbachev) involved in one event — dissolution of the USSR. They died during the 100th anniversary of the USSR. This links them all into one logical category. How to assess the probability of coincidence, when all four individuals who participated in the dissolution of the USSR died in the same year when the population of the former USSR was celebrating 100 years since the formation of the USSR in 1922?

Let's try to get an answer using simple calculations. Suppose the starting point is the year 2007. This year was marked by the death of a key player in the dissolution of the USSR — Boris Yeltsin. Let's also assume that the life period after this is 20 years, during which other participants may die with some probability. But they all die in one year, i.e., 15 years after the death of Yeltsin, on the 100th anniversary of the USSR. If we throw 4 balls into 20 holes and all of them fall into the hole with the number 15, then the probability of such an event is $(1/20)^4 = 0.0000062$. This is an incredibly small probability.

In reality, the probability of death is not a constant ($=1/20$) over 20 years. Statistical observations indicate that the probability of death increases by approximately 10% each year after the age of 65. According to statistical data from the Social Security Administration of the United States government (2020), the values of the probability that a person will die at a certain age are shown in Table 6.4. Members of the Communist Party of the Soviet Union lived in relatively good conditions, comparable to the standard of living in the USA. Therefore, this statistical data should be sufficiently appropriate for the purpose of our calculations.

The probability that all four party leaders will die in one (known) year can be obtained by multiplying four probabilities:

$$0.14 \times 0.14 \times 0.22 \times 0.045 = 0.0002.$$

As before, we will multiply the obtained probability by the number of opportunities for synchronicity (300) from chapter 5.3 The number possibilities for synchronicity. The resulting value equals:

$$300 \times 0.0002 = \mathbf{0.06}.$$

This value (6%) is about the same as the probability of getting "heads" (or "tails") 4 times in a row when flipping a coin. This is the probability of chance creating one synchronization event related to Gorbachev (or one of the other four participants).

Name	Age in	Probability of Death
Burbulis	76	0.045
Gorbachev	92	0.22
Shushkevich	88	0.14
Kravchuk	88	0.14

Table 6.4. *Average death probabilities obtained according to the statistical data from the Social Security Administration of the United States government (2020).*

In these events, the death of Gorbachev, or the celebration of the 100th anniversary of the formation of the USSR itself, was most likely the focal point of synchronicity. This triggered the expectation of the death of other, less significant party leaders who participated in the dissolution of the USSR. However, it should be noted that 6% is not a very small probability, so such events could very well have occurred due to random chance.

6.5 Hitler and Napoleon

If we talk about historical figures, Adolf Hitler (1889 – 1945) and Napoleon Bonaparte (1769 – 1821) can easily be included in the list of the 100 most well-known individuals. Their well-documented biographies share an incredible number of similar elements. Both came to power during the era of a new European democratic republic, and subsequently became known as dictators. Both initiated a series of wars that affected much of Europe's territories. They attempted to conquer the territory of modern Russia and the adjacent Slavic states but failed. This led to the almost complete destruction of their armies and the subsequent fall of their rule.

The number 129 is quite significant for both aforementioned historical figures and for the entire world. This number seems unremarkable at first glance. However, it is associated with some interesting coincidences:

1. Napoleon encountered the French Revolution in 1789. Hitler experienced the German Revolution in 1918 (with a difference of 129 years).
2. Napoleon crowned himself Emperor in 1804. Hitler came to power in 1933 (with a difference of 129 years). Both came to power during the times of a new European democratic republic. Both subsequently became known as dictators.
3. Napoleon entered Vienna in 1812. Hitler entered Vienna in 1941 (with a difference of 129 years).

4. Napoleon attacked Russia in 1812. Hitler attacked the USSR in 1941 (with a difference of 129 years). Note, these attacks also occurred quite close to each other in date (June 24 and June 22).

5. In 1815, Napoleon abdicated the throne and arrived in Jamestown on the island of Saint Helena. Hitler lost the war in 1945 and took his own life (a difference of 130 years). Perhaps this is the only deviation from the number 129. However, it should be noted that Napoleon's fall was not as instantaneous as Hitler's. Arriving in 1816, the new governor Lowe restricted the freedom of the deposed emperor on the island of Saint Helena. From 1816, Napoleon's health began to deteriorate due to a sedentary lifestyle and depression. It can be stated that this year marked the end of his active life (with a difference of 129 years). Here we allow for some margin of error, as they both ended their lives in quite different ways. Napoleon died in 1821.

We will calculate the probability of occurrences (1) - (4) separated by a span of 129 years. We assume that these four random events must occur over 40 years of active life, starting from age 16 and ending at age 56, when both dictators lost power. Events (1) and (5) set the start and end of the time interval. Section 20.6 Appendix provides the program code for such a calculation. The probability that (any) four random years coincide is:

$$8.8 \times 10^{-6}.$$

We will assume that such similarity in the destinies of the two dictators is related to a single synchronicity instance, as they are all connected with one number. As before, we need to multiply the obtained probability by the number of synchronicity possibilities over a lifetime (300) from chapter 5.3 The number possibilities for synchronicity:

$$300 \times 8.8 \times 10^{-6} = \mathbf{0.0026.}$$

This probability (0.26%) corresponds to the emergence of a synchronicity with one of the dictators due to pure chance.

You might ask again — could there have been a huge number of other important dates in the lives of the two dictators that we ignored because they did not coincide with the number 129? After all, was the number of synchronicity possibilities (300) only assumed for an average person, not politicians of a significant importance? The only thing that comes to mind in relation to these two political figures are wedding dates, and perhaps, some other events (like the dates of assassination attempts on Hitler and Napoleon). However, such events were not as fateful for Europe as events (1) — (5).

Another line of reasoning is also possible: Suppose Hitler noticed some coincidences with Napoleon and, as a person inclined towards mysticism, decided to use the number 129 for his most important decisions. But this is highly unlikely: Repeating the attack on the USSR knowing that Napoleon's campaign against Russia exactly 129 years ago ended in complete failure seems quite foolish.

As I said before, time does not exist for synchronicity. If we conclude that the calculated probabilities for random occurrences are small, then synchronicity has certainly manifested itself. In this example, either Napoleon or Hitler initiated events of synchronicity. If it was Napoleon, then the wave of synchronicity "touched" the times of Nazi Germany, that is, the future. However, it's most probable that there was an "adjustment" of information about the past. The anticipation of tragedy by large masses, or the enthusiasm of Germans for the Führer in the 1930s, caused a revision of historical lines, drawing parallels with Napoleon.

6.6 The Kaiser and the War

Wilhelm II (1859 – 1941) — the last Emperor (Kaiser) of the German Empire and King of Prussia from 1888 to 1918, was one of the main initiators of World War I. On August 1, 1914, Germany declared war on Russia. The pretext was Russia's refusal to comply with the conditions of the German ultimatum to cancel the general military mobilization. The mobilization was introduced in Russia in response to Austria-Hungary declaring war on Russia's ally — Serbia.

Thanks to his speeches and interviews, Wilhelm gained a reputation as a confident militarist. He encouraged the military aims of generals and did not allow for any chances of a peace compromise. After the war was lost by Wilhelm, according to the Treaty of Versailles in 1919, he was declared a war criminal and the main instigator of World War I. He was exiled to the Netherlands, where he continued to collaborate with the Nazis and quite profitably invested in the German military industry. Wilhelm greatly admired Hitler's successes in the first months of World War II and personally congratulated him on the victory over the Netherlands in May 1940.

The total number of victims among the military and civilian populations in World War I was about 40 million people. During the period from 1914 to 1918, more than 30 countries were involved in the war. Many historians call World War I "the mother of all disasters" that befell humanity in the 20th century (Belousov and Manykin 2014). It was one of the causes of the October Revolution in Russia in 1917, which subsequently led to millions of human casualties. From a practical standpoint, World War II was a consequence of World War I, which continued after a twenty-year ceasefire. The total number of victims of World War II was around 60-70 million. It is impossible to find more tragic events in human history that led to such massive destruction and a huge number of deaths.

German Kaiser Wilhelm II held a central role in these events that forever changed Europe and the entire world. He certainly belongs on the list of the top 100 most influential people, although his role in the world tragedy of the 20th century is downplayed compared to Hitler. I do not mean to say that the Kaiser was the sole culprit behind all these events. But it was he who was at their center; he militarized Germany, declared war on Russia, and incited war with his public speeches, without considering even minimal chances for a peaceful resolution of the conflict. Historians note that one of Wilhelm's most striking character traits was his abrupt mood swings and his passion for delivering impromptu emotional and harsh militaristic speeches.

Even before the start of World War I, there were an extraordinary number of predictions (Davies 2018) about the incredible tragedy that humanity was about to face. Propaganda also played a significant role, see <u>Figure 6.6</u>. The result did not take long to materialize. In 1915, several British newspapers reported that a student from Montreal had found "solid" evidence that the Kaiser was an agent of Satan. He discovered a remarkably simple way to associate the word "Kaiser" with the number of the beast, 666. This number is mentioned in the New Testament as the number under which the name of the apocalyptic beast is hidden. At the end of the Book of Revelation (3:18) in the New Testament, where the beast (antichrist) is described, we read:

"Here is wisdom. Let him who has understanding calculate the number of the beast, for the number is that of a man; and his number is 666".

Figure 6.6. *"Friends" (Kaiser Wilhelm and the Devil) British anti-German propaganda postcard from World War I times. Gale & Polden Ltd. 1313. Postmark from 1918.*

In the case of the word "Kaiser", all one has to do is find the ordinal number of each letter in the German alphabet, then add 6 to each number and sum it all up. The result is the number "666". Here is how it looks mathematically:

$$\sum_{i=1}^{N=6} p_i + \text{"6"} = 666$$

where p_i is the position number of each letter in the German alphabet in the word "Kaiser". Note that we are simply adding 6 as symbols, not summing up the numbers in the mathematical sense. Is it really that simple? To make it clear, let's look at this calculation:

- K: 11 + 6 = 116
- A: 1 + 6 = 16
- I: 9 + 6 = 96
- S: 19 + 6 = 196
- E: 5 + 6 = 56
- R: 18 + 6 = 186

  Total Sum = 666

Why should the number 6 be added? Firstly, it cannot go unnoticed that the word "Kaiser" itself has 6 letters. We dedicated chapter 3.10 Remarkable number to the number "6", where we explained why this number is exceptionally significant. This number refers to people with their fall into sin. It is precisely the 6th commandment that warns "Thou shalt not kill", thus bringing symbolism to the history with the Kaiser.

Let's check everything ourselves. In section 20.7 Appendix, we reproduced the calculations shown above. Indeed, the number "666" can be derived from the word "Kaiser" using a basic code and the German alphabet. Then, we will calculate the probability that the number "666" can be associated with a random word consisting of 4 to 12 letters, using 60 most simple methods by assigning letters to their positions in the alphabet, and then performing simple manipulations (addition, subtraction, and so on). The resulting probability is 0.007. A probability of 0.007 of obtaining "666" from a vast number of random words is not very small. However, out of 1,000 random groups of Latin letters, which could somehow be associated with "666", not a single group resembled the word "Kaiser", some different European name, or an English or German known word. To see this word (or a word resembling a name), much larger statistics would be required in the calculation. According to my estimate, the probability of obtaining a name that can be tied to the number "666", using 60 of the simplest methods, is very low, possibly much smaller than

$$7 \times 10^{-6}.$$

Such a small probability is indeed surprising. Recall, in 1915, there were no computers, so finding such a simple algorithm to link the word "Kaiser" with the number "666" appears exceptionally astonishing. Is there any mathematical method by which one could find a simple mathematical formula using the word "Kaiser" and 666? I don't think it's easy. Unfortunately, I couldn't find the original source of the article published in 1915.

Another possibility is that the number "666" in the Bible was a coded word for "emperor", referring to a Roman emperor. Thus, it was a word for "Caesar" (Latin), or according to other historical spelling rules, the word was spelled as "Cayser", "Keisari" or "Caisere". However, none of these lead to 666.

What could this mean? The chance of finding such an algorithm with the number "6" is practically excluded. Is it really a warning from the Bible that such a global tragedy could occur? Was this warning specifically pointing to the German Kaiser?

I think the reality might be even more fantastic. It may not be a prediction, but rather a synchronous change of information about the past relative to 1915. When a vast number of frightened people start looking for someone to blame for events that are occurring, they try to find some simple explanation or algorithm to confirm their worries or expectations. In this case, the main idea is that such an algorithm must contain the number of "human sin" — "6". Then, an alphabet and this simple algorithm are used to find 6 position numbers in this alphabet. It must lead to some "word" with 6 letters. The combination of the letters "Kaiser" emerged. It means absolutely nothing, as the word "emperor" was spelled somewhat differently before 1915. But then, a synchronicity occurs. It assigns the letter group "Kaiser" to the word "emperor". Now, "Kaiser" began to be used for German emperors prior to 1915. Such a synchronicity mechanism leads to a simple connection of Emperor Wilhelm II with the number "666" from the Bible, as many people wanted. In this explanation, it's not necessary to deal with probabilities of the order of 10^{-6}. The symbolic word "Kaiser" was created in 1915. We will examine this example in detail, using the symbolism in chapter 16.4 Symbols and forms of ideas.

It must be acknowledged that my explanation for obtaining the number "666" is pure speculation in this historical example, but doesn't it intrigue you, nonetheless? Such interpretation is consistent with an event of synchronicity, which is not subject to the flow of time. Note that changing information about the past does not necessarily mean changing the past itself, which is inaccessible to us. We will return to this hypothesis when we discuss quantum mechanics in chapter 8.2 Nonlocality. In the microworld, this concept is called "retrocausality". If the human brain is somehow connected to quantum-mechanical effects (Penrose 1989), then similar phenomena could manifest in society during moments of the greatest psychological stresses. Admittedly, this is purely speculative at this point, but is it not a good explanation?

As we did before, let's multiply the obtained probability of 7×10^{-6} by the number of synchronicity opportunities (300) from chapter 5.3 The number possibilities for synchronicity:

$$300 \times 7 \times 10^{-6} = \mathbf{0.0021.}$$

This value (0.21%) represents the probability for blind chance to create at least one synchronicity associated with Wilhelm II.

6.7 Again about wars

Searches for a symbolic connection between the First and Second World Wars led to an interesting observation that has become a topic of discussion on the internet. As is often the case, numbers are the most suitable way to establish such a connection between events. It turned out that one can take the dates of the beginning of the First and Second World Wars, break them into groups of two numbers and then add them. The resulting number is 68. See:

- Start of the First World War: 07-28-1914 leads to
  $07+28+19+14 = 68$
- Start of the Second World War: 09-01-1939 leads to
  $09+01+19+39 = 68$

Here I use the American date format (MM-DD-YY), in which the month comes first, followed by the day and year. The result will be the same for the European style of writing dates (DD-MM-YY).

What is the probability that the number 68 (or any number) coincided by chance? To calculate this, one needs to establish the range of possible numbers (or the number of possibilities). The maximum possible number can be defined as 12-31-1999 or $12+31+19+99=161$. The minimum number can be set by the date 01-01-1919 or $1+1+19+19 = 40$. Therefore, the probability that a fixed number (for example, 68) for the Second World War will arise randomly is $1/(161 - 40) = 0.0083$. In fact, the more precise value is somewhat larger since we haven't taken into account that the number 68 can be obtained by permuting numbers in the groups of two numbers. But I'm not aiming for good precision in this discussion.

It should be noted here that we have not exhausted all possibilities. I think there are about 10 simple algorithms for obtaining a certain positive number from dates. For example, numbers can be multiplied or added symbolically (i.e., "19" + "19" = "1919"). We will not consider complex algorithms. Therefore, a more realistic probability of the appearance of the number 68 using several simple algorithms is

$$10 \times 0.0083 = 0.0826$$

or about 8.3%. This probability is not small enough to exclude the possibility of chance.

As I mentioned earlier, it is precisely in moments of global political shifts and wars that people seek to find explanations and perpetrators of tragic events. For example, during the outbreak of the conflict between Russia and Ukraine in 2022, a large mass of people in the post-Soviet space saw a direct analogy of this event with the Second World War. This new confrontation looked like a typical conflict between East and West due to cultural differences. The inevitable Third World War was on the horizon, logically following from the Cold War and the Second World War, just as the Second World War followed from the First. All logical parallels between the new impending war and the Second World War were obvious. Of course, there had to be some sign linking these events. It was this moment of imminent tragedy that prompted people to seek connections between historical events to find the causes of the upcoming suffering and tragedy.

The results of such searches were not long in coming. It turned out that the beginning of the official Russian invasion of Ukraine (02-24-22) gives exactly the number 68:

• Russian-Ukrainian conflict: 02-24-2022 leads to
  02+24+20+22 = 68

Of course, such a surprising coincidence led to active discussion in the online space. The probability that all three wars are connected by some simple algorithm and some number (such as 68) by chance will be:

$$0.0826 \times 0.0826 = 0.0068$$

or 0.68%. This is a small probability, roughly corresponding to the chance of getting "heads" seven times in a row when flipping a coin. It is hard not to notice that the number 68 seems to "dance" around these three events, although such a coincidence exists only until the probability 0.0083 is increased by the number (10) of possible algorithms, which is poorly defined. Namely, the fact that we took 10 algorithms is simply a coincidence, as this number may be somewhat larger or smaller. We will see something similar in chapter <u>10.1 Nostradamus</u>.

The search for parallels with World War II did not end there. During the onset of active military actions in 2022, the peoples inhabiting Russia saw a follower of Nazism precisely in the Ukrainian president. And again, a strange coincidence did not take long to manifest. It turned out that the wife of the Ukrainian president (Zelenska) was born on the same day of the year, February 6th, as Eva Braun (1912 – 1945), Adolf Hitler's longtime companion. Perhaps, it is not very surprising. As we discussed in chapter <u>4.3 The birthday problem</u>, it only takes an average of 23 people to find a pair with the same date.

However, in this case, it is precisely the semantic connection with the same factor (the companion of a person whom people associate with belonging to the same category of views) that is interesting. It does not matter at all whether the Ukrainian president supported Nazism within Ukraine or not. The primary factor was the widespread belief among Russia's numerous ethnic groups that the nationalist government on their border posed a significant threat. By all its characteristics, this new conflict was a continuation of the wars started by the

Kaiser and Hitler. The willingness to fight against Russia for European ideals and the refusal to oppose nationalist sentiments within Ukraine were enough to forever link the Ukrainian leadership with the symbol of Nazism in the minds of many.

So, let's return to the connection between Zelenska and Braun. Both women posed with their partners on the cover of the British magazine Vogue. As we said previously, the number 6 in their birthdays also has significance in religious numerology (relating to human fallibility and the commandment "thou shalt not kill"). The 66-year difference (!) between the births of these two women further decreases the likelihood of a coincidence and increases people's confidence in their guesses that it is some sort of sign, and which side of the conflict to take. This rare coincidence quickly spread across major social networks.

Let's estimate the probability of this unusual occurrence. We are dealing with four people (two pairs) and 8 numbers. These are the birth and death dates for the first pair (Hitler and Braun) and two birth dates for the second pair. The probability that the birth dates of the second pair will coincide with the birth (or death) dates of the first pair is $4/365 = 0.01$. So, this is 1%. This number does not take into account that the birthday itself fell exactly on the 6th (or 16th), that is, on some expected number with the digit 6. Such a probability is approximately $2/30 = 0.066$, assuming there are 30 days in a month. Here, we do not necessarily have to be precise, that is, it does not matter whether there are 31 or 29 days. Then, consider that the difference between birthdays is 66. This matched the expectation to find a 6. Such expectations could also be related to 6, 16, if everything happened within a century. This leads to a probability of $3/100 = 0.03$. Thus, we obtain the probability that one of the birthdays coincides, and at the same time, numbers with a 6 appear for the day of the month and for the difference between the births:

$$0.01 \times 0.066 \times 0.3 = 0.0002.$$

By multiplying the obtained probability by the number of synchronicity opportunities (300), we get 6% for a random chance to create some synchronicity in the lives of these two women.

Perhaps, this is another example of editing information about the past. In this case, this phenomenon could be due to the expectations and fears of large masses of people who started looking for coincidences in the present, leading to changes in the related historic events. However, in this example, the probability that it's just a coincidence is not as small as in the case with the Kaiser. In this case, if it was an event of synchronicity, the birthday of one of the women was subject to revision.

6.8 From my experience

The main principle of my approach is to consider only the incredible events of well-known people, as their biographies are well-documented. The events themselves must be quite significant for their lives, that is, related to birth, marriage, death, or another pivotal moment in their lives. In the case of famous people, such events can also be momentous for the entire world.

I do not belong to the category of famous people marked by me for statistical experiments. Nevertheless, as the author of this book, I want to talk about my personal observations in the moments of my life that were critical for me. What I experienced over 20 days during a period of emotional stress is impossible to explain rationally. It was the observations made during that period that were the reason for writing this book.

I began planning this book right after I was struck by grief — my mother died, and I had to travel to her funeral, facing many difficulties to reach the geographic center of Europe — Minsk, the capital of Belarus. When my mother was put into an induced coma after clinical death, my sister and I arranged that she would go for ten days first, and then I would follow for ten days with a two-day overlap. During these two days, my sister was to hand over the care of our mother to me. With this plan, we booked the plane tickets. My mother passed away, but in such a way that the funeral was precisely during those two days when my sister and I were together. Had she died earlier or later, it's almost certain we wouldn't have been together to bury her.

I had to go to Minsk. This became the beginning of a series of coincidences that I simply cannot explain logically right now. I must say that the trip was not too easy. At that time, NATO member countries did everything possible to prevent anyone from entering Belarus by plane or train. There were no flights from these countries to Belarus. The small Lithuanian airport in Vilnius served flights from the military alliance. The new Minsk airport provided services to countries of the former USSR (except Ukraine and the Baltics), the Global South and Asia. Flying from America, I had to fly to Vilnius and then cross the border by bus into Belarus.

Event 1

My mother was buried in the 78th spot in the columbarium, exactly 3 months before she would have turned 78. Her father was 78 years old when he died. What is the probability of such a coincidence?

Usually, a person is associated with about 10 significant numbers, each date consisting of 2 numbers (day of the year from 1 to 365, and the year itself). If the number 78 comes up (the number of the place in a columbarium), what is the probability that a person's age will also be 78 and the age of their father when they died will also be 78?

The main problem in such estimates is that the numbers significant to a person are not evenly distributed. Smaller values occur much more frequently than larger values. Days, months and hours are distributed between 1 and 30. The number 78 is quite rare compared to the number 12, which occurs in dates. Conservatively, we assume that all significant numbers are evenly distributed between 1 and 30. If you have a 30-sided die, what is the probability that a certain number will come up 3 times in a row? Let's calculate this:

$$(1/30) \times (1/30) = 0.0011$$

or 0.11%. It is a small probability, even considering the conservatism in our reasoning: We assumed a range of values from 1 to 30, not in a range from 1 to any larger value (like 78). We did not consider the range from 1 to 78, as the probability that a person will die young is quite low. In this case, we simply assumed that we are dealing with an equal probability (1/30) in the age range of 48 to 78.

Event 2

My mother was dying within 16 days of being picked up by an ambulance on November 4, 2023. My sister, having arrived in Minsk a little later, noticed that during the 21 days between November 9 and December 1, there were 4 days when the sun came out. This was also noted by all the friends and relatives present at the funeral. On all other days, there were thick clouds. During these days, there were 4 significant events: the day of death (November 20th), the funeral (November 24th), the commemoration or 9 days after death (November 28th), and the moment of burial (December 1st). Orthodox belief holds that the soul of the deceased remains on Earth for 9 days after death, only then ascending to heaven. What is the probability of the sun appearing precisely on the days when important events occurred, considering that the appearance of the sun happens randomly, as is often the case in this part of Europe?

To calculate the probability of such a coincidence, we need a computer code. The program is shown in section 20.8 Appendix. The obtained probability is:

0.00017.

This is a small probability for an event that occurred randomly.

This story is interesting because when the sun appeared after many days of bad weather and low clouds (November 20th), my sister interpreted this day as a sign of my mother's recovery. However, contrary to her expectations, she was informed that our mother had passed away. Our mother had spent 10 days in a medical coma. The subsequent clearing of the weather and observation of the sun during the three significant events related to the funeral and memorial services were confirmed by many relatives with whom I discussed this phenomenon.

Certainly, there could have been other signs, such as a cat crossing the road, snowfall, and so on. However, these events cannot be considered sufficiently important, and they are not capable of significantly increasing the number of possibilities. Even if a dozen of such unfulfilled signs were to accumulate, it would not alter the incredibly small probability of what happened.

One of the key questions in this example is whether we can realistically consider scenarios where all the clouds over a vast city change due to the fate of one person. There is nothing surprising in this for those who are familiar with the weather in late autumn at this latitude of the European continent. The sun often appears for a brief period and then suddenly disappears; the location where the sun appears also changes significantly within a fairly short distance.

Event 3

When choosing an urn for burial, we settled on a light-colored square urn from a funeral service store. After the funeral, we decided to redo the plaque for my father, who was buried 7 years ago. When the funeral service workers broke the slab covering the father's ashes, it turned out that he had exactly the same urn. This was confirmed by my sister Natasha, who was present at the burial. Neither my sister nor I knew what kind of urn my father had, since we were not able to visit the burial of his urn 7 years ago.

Let's calculate the probability of such a coincidence. There were 3 urn colors in the store (white, blue, black), — probability 1/3. The number of urn shapes with different designs was around 10. Therefore, the resulting probability is 1/10. The overall probability that the urn was chosen correctly:

$$(1/3) \times 0.1 = 0.03.$$

This is also a quite small probability, which intrigued me.

How to explain this

The probability of each of these coincidences, occurring at the moment of high emotional intensity, is quite small for pure chance. The combined probability of the simultaneous occurrence of these three events over 21 days, related to a single logically-connected event, is obtained by multiplying the numerical values obtained earlier:

$$0.001 \times 0.00017 \times 0.03 = 5.1 \times 10^{-9}.$$

Since all these events correspond to one significant event (the death of a loved one), we will consider this probability to correspond to one synchronicity event. It is inconceivably small that such a combination of events would occur randomly over a short period. In my life, I have never experienced so many coincidences related to a single topic.

As before, we will multiply the obtained probability by the number of synchronicity possibilities (300) from chapter 5.3 The number possibilities for synchronicity

$$300 \times 5.1 \times 10^{-9} = 1.5 \times 10^{-6}.$$

This is a probability for a random chance to create a synchronicity instance in my life, or in the life of my sister (since we observed such events together). Of course, with such a small probability, randomness is unlikely to have played a role here.

As I mentioned, there are many improbable events in the world, considering the number of people living on Earth. I do not consider myself significant enough, so my synchronicity accident will not be considered in the conclusion of this chapter. Nevertheless, the calculated probability is small even for 10 million people. All these events occurred in an incredibly short and emotionally critical moment. As we have repeatedly discussed, research shows that synchronicity effects occur precisely in such life-changing moments.

These coincidences operated on a significant circumstance (the death of a loved one) and revealed themselves with the help of significant and well-noticed signs for all participants in these events. These were numbers and natural phenomena, such as weather. Therefore, even if there are any other possibilities for signs related to my mother's funeral, their number would be very insignificant, and cannot change the outcome for the calculation. Thus, the obtained small probability for blind chance will not increase significantly if all unaccounted possibilities for the appearance of signs are included in our calculations.

The combination of such coincidences, from my point of view, can only mean one thing — intervention in the rational course of events. These were symbols that were meant to be deciphered by me only. I have already mentioned that my mother was not a very religious person, nor was our family. But her soul knew that, as a scientist who had spent my entire life with computers, numbers and probabilities, I would be able to decipher such signs and show that they are indeed real and cannot arise randomly.

Such events occur when there is an interacting system of people and the surrounding elements of nature. These are not just some external phenomena sent by someone. We are part of such a system, and we can perceive them because our consciousness has common origins. We not only perceive such phenomena, but also actively participate in them. These incidents may be initiated by chance, but the observers must resonate with them for better perception and amplification, elevating them to a level where pure chance ceases to be a good explanation.

It is quite possible that these rare events occurring during the funeral were meant to convey a message to us. They suggest the presence of a reality beyond our material existence. It needs to be understood. These events were my mother's last will and testament, and an instruction to me and my sister to share it after her death. And I took her signs as an invitation to write this book. I completed writing this book three months after the funeral to ensure that I could capture all the details while they were still fresh in my memory.

6.9 Nothing is accidental

I have presented 6 (not counting my own) instances of statistically improbable circumstances that have occurred to people who fall within the group of the 100 most known people. My goal from the start was to not deal with the 8 billion people with whom something unusual can happen at some time and place. Instead, we narrowed down our thought experiment to a rather small statistical sample of people. In addition, we only considered the most significant events in their lives.

Of course, such coincidental events can happen to anyone. However, finding evidence of this phenomenon will be significantly more difficult when dealing with the Earth's multi-billion population. We created a "filter" to reduce the number of potential possibilities (i.e. the size of the statistical sample). This was achieved by focusing our attention on well-documented historical cases.

In simple terms, we discovered a group, within which 6 people tossed a coin 5–6 times in a row, each time landing on "heads" (or "tails"). This is an exceptionally rare occurrence. Generally speaking, getting "heads" 5 times out of 5 coin tosses is an improbable event even for 2 people in a group of 100.

What is the probability that all these 6 historic events indeed happened? Since all such events are fully independent, we will obtain it by multiplying all 6 probabilities, highlighted in bold, in the six previous chapters 6.1 – 6.6:

$$7.5 \times 10^{-4} \times 0.0187 \times 0.024 \times 0.06 \times 0.0026 \times 0.0021 = \mathbf{1.1 \times 10^{-13}}.$$

This is an incredibly small probability for a group of pre-selected people in our thought experiment. Note that in the case of the Kaiser of Germany, we took an exceptionally conservative probability.

It is evident that this is not the probability of the emergence of the initial starting point of synchronicity, as discussed in 4.2 Probabilities of synchronicity, where we excluded the possibility of the very source of synchronous events. In these six examples, it was difficult to establish the reason why synchronicity began to manifest itself.

I'll remind you that we applied the following approach in this analysis:

- Used a group of well-known individuals with well-documented biographies (about 100 people). We did not provide a complete list of these individuals to avoid unnecessary disputes. However, I believe the people used in our examples will not raise objections for most of us.
- Considered 10 significant events for each individual (birth, death, marriage, graduation, and so on). These have already been accounted for in our calculations.
- Assumed about 30 situation-connections with each event out of the 10 important events.

The only thing we did not account for is that we are dealing with a group of 100 people. Thus, the numerical value of the probability that we expect for the six coincidences is:

$$1.1 \times 10^{-13} \times 100 = 1.1 \times 10^{-11}$$

That is, we are still dealing with an astronomically small probability for our small sample of people and events.

Even if there were some errors in our calculations, I do not believe such an error could significantly alter our conclusions. In most cases, we considered quite conservative scenarios and their probabilities. Of course, if we consider 8 billion people, then we get a probability of 0.00088. This is a much larger value, and it is not astronomically small compared to a group of 100 people.

How should we interpret the probability of 1.1×10^{-11}? It means that there is only one world out of 1.1×10^{11} worlds (assuming their similarity) where a group of pre-selected individuals with identified synchronicity exists. More precisely, this probability represents finding a group of 100 people where 6 individuals exhibited synchronicity purely by chance. In the previous analogy with the coin, this is equivalent to discovering a group of 100 people where six individuals tossed a coin 5-6 times, and it happened to land on heads every time. Notably, we didn't even try hard to find these individuals, as their names are well-known to many.

How is the emergence of this group of individuals possible in a world governed solely by chance? Statistically, such an event should not be observable. This raises questions about this world. Could we have uncovered evidence that significant situations in our life are influenced by external factors and do not occur merely by blind chance? Of course, the term "evidence" is used here assuming my arguments make sense for the reader.

Unlike the cases presented in Carl Jung's book (Jung 1973), which justifies synchronicity using his own experiences and the observations of his patients, our calculation is based on a strict principle that narrows down the statistical sample to a small number of people. These people became famous not because something odd happened to them, but because they were already quite well-known before these improbable events occurred. We also provide precise numerical calculations that can easily be replicated with basic skills in programming.

We have already mentioned that we can observe the adjustment of parameters in the physical laws for the existence of our Universe. Such a phenomenon does not seem random. The formation of DNA and cells required by complex life could also be connected to some influence that disrupted the rules of the emergence (more precisely, the non-emergence) of information from matter. If all these events cannot occur by mere chance, then there must be some mechanism leading to such phenomena. Could our world have been somehow changed or edited? Of course, this intervention must be intelligent. In chapter <u>16 Beyond this reality</u>, we will attempt to explain such coincidences.

6.10 Riding the wave of coincidences

If strange coincidences have not yet happened to you, here's my advice — just wait. Synchronicity often arises in waves. At some point, you too may find yourself at the "crest" of such a wave. Waves compress statistically improbable events into groups by decompressing them in other situations.

As I've said before, it was specifically a group of synchronicity events in November 2023 that inspired me to write this book. I wrote it after work and during the rare moments I could find on weekends. By March 10, 2024, it was completed. I firmly decided then that I would not add anything more to the text. The initial subtitle of the draft was "Computational Arguments in Favor of the Existence of God". But in February 2024, I revised it to "The Incredible Reality Beyond This World". I thought that this might add a certain touch of mystery and spark more interest among readers. The hope that the reader might infer the presence of God after reading this book was also taken into account. And, perhaps, such a subtitle will attract the attention of atheists and those who think our world is a computer simulation alike.

But then an event occurred which so astonished me that I decided to supplement the draft with this chapter. I am a rare visitor to the social media network X/Twitter (blocked in Russia). Once a week, I posted messages about a project related to the Encyclosphere, to which I devoted my free time. The Encyclosphere is somewhat similar to Wikipedia. It combines dozens of different encyclopedias together and "decentralizes" them. In simple terms, encyclopedia articles get their own "life" in the form of files, which can be exchanged and downloaded to computers. This project was developed by a non-profit organization called the Knowledge Standards Foundation. Dr. Larry Sanger, a co-founder and former chief editor of Wikipedia, was the president of this organization. Since 2021 we have been meeting with Larry once a week via video to discuss technical issues of the Encyclosphere. The topic of God never came up in our discussions. Mostly, we worked on the software aspects of this project. It's worth noting that Larry maintains a blog about Christianity on X/Twitter, and I often saw his posts on religious topics. But I was not very interested in religion, considering it just a cultural "layer" handed down to us from our ancestors.

On March 10, 2024, I read a message from Larry on X/Twitter stating that the draft of his book "God Exists" (Sanger 2024) was finished, and he could send it to anyone who wanted to read it. By that time, my book was already written, and I was waiting for the proofreading of my Russian version. I had never told Larry about the book I was working on, which also suggests that God is one of the most plausible explanations for our life. I immediately responded to Larry's message: "Larry, what is the probability that two people who meet every week to discuss technical issues of a joint project would write two books on the same topic, without ever mentioning them in their conversations, and moreover, finish writing them at approximately the same time?"

On that day, we exchanged drafts. My book was in Russian, his was in English. Both books were dedicated to our parents. Many sections were quite similar: The origin of the Universe, the emergence of life and the formation of social morality. Larry's book explored philosophical aspects, while mine focused on coincidences, scientific contradictions and various inconsistencies in understanding the world. Naturally, his writing style was very different from mine. He wrote in a fairly strict academic style typical of a philosopher. My style was more concise, as is customary in articles dedicated to natural sciences.

Calculating the probability of such a coincidence, where two people write books on the same topic without ever mentioning them during a long period of personal communication and finishing them in the same year and month, is challenging. I think the probability of a random coincidence is astronomically small. I hesitate to give a numerical estimate for such a coincidence. In the past, neither Larry nor I had written books about God and religion. I had never written books for a general audience. Larry wrote articles and books about social aspects and politics. As I understood from his draft, he was agnostic until 2020. I was also agnostic about God until 2023, when I experienced events of synchronicity (see chapter 6.8 From my experience). Even if we assume that Larry had a professional reason to write his book, expecting a similar manuscript from me was entirely out of the question, which he admitted.

6.11 Signs of the material world

Amidst the endless whirlwind of interconnected events through various causes and effects, and random circumstances that manifest without known reasons, there are coincidences between important moments in our lives and phenomena that are understood only by those who observe them. It is difficult to establish their existence using scientific methods, as they are fundamentally subjective and cannot be easily verified by independent experiments.

We mentioned earlier that Wolfgang Pauli (1900 – 1958) was one of the most brilliant physicists of the 20th century. He predicted the existence of the neutrino — one of the most fundamental particles — and later received the Nobel Prize. Pauli was a theorist, which is often perceived as being directly opposite to experimentalists, who deal with equipment and setups. He was also known for the fact that sometimes, when he entered a room or an experimental laboratory, something extraordinary may happen. Experiments would fail because the equipment would start to malfunction. Colleagues jokingly called this the "Pauli Effect". Perhaps all of this is nothing more than coincidences and circumstances, as there is no solid evidence proving that people can somehow influence instruments. However, some members of the scientific community, including Pauli himself, believed that it was real. But how could scientists, committed to objective and verifiable laws of nature, believe in such an effect without any scientific proof?

By the way, there are many examples to the contrary, where a malfunctioning electrical device starts working as soon as a specialist approaches it, or a computer begins performing calculations correctly as soon as an experienced programmer examines and studies the problem causing the software glitch.

163

The connection between the world of people and the world of physical objects has always held great significance in the minds of many. People believe in this connection even without any solid scientific basis. However, no one has yet succeeded in bringing this belief to the level of objective and easily verifiable knowledge. Official science cannot establish such a connection.

However, the correlation between significant life events and phenomena in the surrounding world was never doubted in the distant past. These correlations were called "signs", and they played a crucial role in decision-making by our ancestors for centuries. The interpretation of signs and various omens in the physical world, pointing to future events, was first developed in ancient Mesopotamia (around the 4th millennium BCE). Collections of omens, interpreting signs both in the sky and on the earth, were first recorded during the Old Babylonian period (early 2nd millennium BCE). Since then, the connections between events in the human world and phenomena in the physical world have been the most common way to understand the essence of what is happening.

I decided to write this chapter in August 2024, after the first edition of my book had already been published. I added it solely to explain how signs can arise, how to interpret them, and how to assign appropriate probabilities to their occurrences. I must say that I have never had the ability to interpret signs in advance, before events have actually happened. I simply never took an interest in such literature, considering it "rejected" knowledge by scientific methods. My ability to foresee events using signs is absolutely nonexistent. However, as a physicist, I am capable of finding connections between natural phenomena and interpreting them within the framework of a scientific hypothesis or theory. This, after all, is characteristic of all scientists who observe the phenomena of nature.

So, in August 2024, I traveled once again to Minsk from Chicago for a week, spending about two days on the journey each way. I had two goals: to officially transfer our family's summer house and apartment to my sister after the death of our parents. I decided it would be very timely to do this in the summer, during my vacation.

As I prepared for this trip, I anticipated the appearance of synchronicity. Having written this book, I was certain that something extraordinary was bound to happen — just as it did during my previous trip when I had to travel for my mother's funeral. But I had no idea what or how it might occur. However, I knew that coincidences often happen during moments of emotional intensity, as I have demonstrated multiple times in my book. These coincidences must involve significant events in the lives of people, semantically related natural phenomena and logically connected numbers.

Here, I need to make a brief digression. Our summer house near Minsk, which my sister and I decided to sell, was built over the course of 40 years. My grandfather started its construction back in the 1980s. Since then, it had gradually expanded with new rooms and additions. My father and I worked on the house mainly on weekends. I helped build the sauna, which is located right next to the house. Three of my uncles and a large number of relatives and friends also contributed to the construction. The summer house was a gathering place for our family for decades. Naturally, this gave the place a strong emotional significance. This house held the spirit of our family. Parting with it was undoubtedly a deeply emotional experience.

One weekend after arriving in Minsk, my sister and I went to the summer house to get rid of unnecessary things and prepare it for sale. The weather was beautiful, with the sun shining brightly. For me, this was my last visit. After we had cleaned everything up, we approached the front door to lock it. For me, it was for the last time. Our neighbors came over to say goodbye. And then, it started to rain! During my entire stay at the summer house and in Minsk, the weather had been beautiful. To our surprise, this was the only time I saw rain before leaving the house.

My father often said, perhaps jokingly, that it tends to rain when you say goodbye or leave a place close to your heart. This phrase stuck with me because we had experienced several "rainy" farewells in the past. I always suspected that it was just a joke, spoken by my father with a serious expression. Naturally, I never connected such natural phenomena with moments of emotional experience. I simply didn't believe in it. Of course, I knew that in literature and art, rain is often associated with sadness, the cyclical nature of life, and symbolizes change. Since ancient times, the image of rain falling from the sky has been compared to tears falling from the eyes of a crying person. This comparison creates a powerful connection between rain and human emotions.

So, I locked the summer house. Tears appeared on my sister's face, and it began to rain heavily. This downpour was quite inconvenient, as we needed to walk to the bus stop. Our neighbors, standing nearby, suggested that the rain would pass quickly — the dark cloud looked rather isolated.

But then it struck me. I thought that this might be a significant sign of synchronicity. If this rain was a manifestation of sorrow over parting with this place, then it should stop immediately. We had to walk about 15 minutes from the summer house to the bus stop, with the bus departing in 20-30 minutes. If this was just an ordinary shower or a random coincidence with an inanimate natural phenomenon, then there would be nothing to prevent the rain from soaking us while we walked to the bus or waited at the stop. However, if this event held some deeper meaning connected to our farewell, and was in harmony with our emotional state, the rain should stop quickly, allowing us to board the bus dry. The downpour and the moment of parting should be synchronized, but the rain continuing after the farewell wouldn't make sense.

The rain drizzled and stopped after about 10 minutes. The sun came out, and we successfully made our way to the bus dry. We didn't encounter any more rain, neither on the way to Minsk nor in the city itself.

As I mentioned, thunderstorms and rain in this part of Europe are local phenomena. The climate here is not continental. Rain in Belarus tends to be concentrated in small areas. It can rain in one spot while the sun shines just a few kilometers away. The wind and low cloud cover, consisting of scattered cumulus clouds, and the peculiar instability in the atmosphere contrast sharply with the weather in central North America and Russia, where rain can cover vast areas of landscape. My sister noted that the clouds here are also unusual — they are very low, often spaced at regular intervals, and move quickly.

We can see that this case exhibits all the characteristics of synchronicity. A significant event with an emotional surge coincided with a notable natural phenomenon — the rain. And it wasn't just any natural phenomenon; it had a pre-existing meaning associated with it, namely sadness or crying.

Can we estimate the probability that this case is merely a random coincidence? Let's try to make such an assessment. We're interested in the number of possible opportunities. I hadn't seen rain for 5 days before the trip to the summer house (and several days after). The emotional surge related to closing the summer house lasted about 10 minutes. The burst of sadness associated with the dacha closing lasted about 10 minutes. Five days have

167

$$5 \text{ (days)} \times 24 \text{ (hours)} \times 6 = 720$$

opportunities for precipitation. Here 6 means the number of 10-minute intervals in one hour. Thus, the probability of encountering rain during the specific 10-minute interval is approximately $1/720 = 0.0013$. Such a low probability suggests that the event may not have occurred purely by chance. Of course, if it had been strong wind rather than rain, it likely wouldn't have caught my attention. Here, a natural phenomenon clearly participated in the event and was linked to a specific occurrence and had meaning. Of course, rain was happening in other parts of the area, but we were witnesses to it during a particular place and time — the moment of sadness.

I do not exclude that other natural phenomena could be perceived by us with a similar meaning — such as sadness. For example, a melancholic melody or unusual bird song might have had a similar effect. To be thorough, we can multiply 0.0013 by 5 possible coincidences with other important and rare phenomena that could be perceived as signs of sadness. Thus, the probability that the moment of parting was "timed" with some significant phenomenon symbolizing sadness would be $0.0013 \times 5 = 0.0065$.

If you believe that only nature participates in the events of your life, as most of our ancestors did, this would be partially true. In reality, it is possible that the entire structure of our surrounding reality seems to be modified based on our emotional involvement.

Here's another example from the same trip. The second purpose of my stay in Minsk was to sign a document — the deed of gift transferring the apartment to my sister. We spent the entire week gathering documents, and after endless visits to various offices, we finally arrived at the department where we needed to sign the papers. Any inaccuracy in the paperwork could have derailed the deal, which would have been a catastrophe since I was leaving for Chicago the next day. This event was also emotionally charged for me, as I was relinquishing the apartment where I had grown up. It was the only property I had in Minsk.

So, with a folder of documents gathered with incredible effort, we arrived at the department where the transaction was to take place. I approached the ticket dispenser and pressed the button to get a queue ticket. The ticket had the number 69. Almost immediately, the red number 69 lit up on the office door where we were supposed to enter. My sister and I exchanged glances and understood everything. There was no doubt — the transaction would be successful. And so, it was. All the paperwork was correctly assembled, and the deal was completed.

You might wonder why the number 69? The reason is that my birthday consists of two blocks of the number 69. The ticket dispenser at that location issued numbers up to 200, as I later checked. The probability of receiving the specific number 69 was 1/200, or 0.005. I never associated the numbers 6 or 9 individually with myself. Naturally, there could have been other numbers that we might have interpreted as coincidences, such as the number of the apartment I was transferring or another number related to my sister's birthday. Even if there were around five such numbers with similar significance, the maximum probability of encountering one of these numbers would be around 2.5% ($0.005 \times 5 = 0.025$). Still, this is a small probability for a mere coincidence.

As we see, this case is not about nature per se. It was simply a mechanism of the ticket dispenser that printed a number recognizable to us. The structure of the observed world appears to be arranged in a way that it conveys a meaning we can comprehend. Whether this is nature or mechanisms created by humans is irrelevant. The material environment seems to be subject to some form of adjustment effect.

In simpler terms, two critically important events for people were synchronized with external physical phenomena.

Certainly, this trip was not so significant in my life as to rank among the top 10 most important events. What I want to convey is that the journey, marked by two key circumstances, was accompanied by synchronicity. These two events were independent of each other and had probabilities of 0.0065 (the last day at the summer house) and 0.025 (the final signing of the apartment documents). The role of random chance, which might have played a part, has a probability equal to the product of these two small numbers ($0.0065 \times 0.025 = 1.6 \times 10^{-4}$). This probability is approximately equivalent to flipping a coin and getting heads 13 times in a row.

One could also pose the question this way: how many trips similar to this one would it take for random events in the physical world to align meaningfully with the circumstances I experienced, assuming such phenomena occur purely by chance? The answer would be around 5,000 trips. Of course, I did not have that many attempts.

I wrote this section while on a bus traveling from Minsk to Vilnius. Outside the window, endless queues of trailers loomed, waiting for customs control for up to five days at the political border of Europe. Behind me lay Belarus and the vast expanses of Russia and the rest of Eurasia. Ahead were the suffocating Baltic countries, cut off from its history by European Union barriers and endless customs checks.

7 Time, space and their absence

7.1 Time

For people and the world around us, time is a quite compre-hensible quantity: It defines the sequence of events. The flow of time, or the arrow of time, moves from the past into the present, and then into the future. The arrow of time itself is a consequence of the in-crease in entropy, or the degree of disorder and uncertainty in the ex-panding Universe after the Big Bang. As we discussed in chapter 3.1 Entropy, the higher the disorder of some system or data record, the greater the entropy. In relation to this Universe, the more chaotic the

171

movement of material particles that make up the world around us, the greater the entropy.

However, the fundamental laws of nature do not depend on the direction of time. Most of them are reversible chronologically. For example, we can calculate the basic equations of physics backward in time just as easily as forward in time. This means that theories with causality forward in time must also have causality backward in time. There is a direct analogy with the coincidences we discussed earlier.

We must not forget that the laws of nature (gravity, magnetism, etc.) set nothing in motion. For them, there is no time. But if we create a computer model of the world or a game based on our world, we have the opportunity to "animate" these laws. Everything will come into motion. If such changes occur with sufficient speed, then the sequence of images constructed according to the laws of nature will seem like a moving picture or film. Why, then, is the future so different from the past in our world? The origin of this arrow of time has puzzled scientists and philosophers for over a century and remains one of the fundamental problems of modern physics.

In our examples about coincidences in chapter 6 Examples of significant coincidences time and causality play no significant role. Space also does not play a role — events can be arbitrarily separated in space, just as they can be in time. Coincidences occur within the realm of significant circumstances, names, symbols, and numbers. They manifest at moments when people experience pivotal situations. The numerical space is "tuned" to human perception, i.e., to the decimal system (we will discuss this later), whereas time is used only for defining numbers (such as the day or year).

In these examples, we can easily turn time backwards, and the examples themselves will not change. The same applies to the probabilities we calculated. Clearly, the reason for the coincidences between the events involving Kennedy and Lincoln in <u>6.1 Abraham Lincoln and John Kennedy</u> is not that Lincoln predestined what would happen to Kennedy. That would be unfair to Kennedy and the future. Here, we are dealing with a situation where unusual events have no arrow of time at all. What happened to Kennedy, in a sense, also affects the events with Lincoln. Or, to put it differently, events that have occurred exist equally regardless of their sequence. Is it possible that the causes of such occurrences are not located in the time and space of this material world?

7.2 The Lawgiver

Have you ever wondered where the laws of nature come from? This question has always been central to the greatest minds of humanity. Many philosophers, mathematicians and physicists tend to believe that the laws were given to the world externally, and that we must discover them. In simpler terms, they are not a subjective reality created by our brain.

Let's clarify this. Science works as follows: It first discovers the laws of nature using observations and experiments. Then, science uses this knowledge to search for further explanations. But the laws of nature are nothing but our description of how nature functions. Surely, they cannot be created by nature itself. Just as the description of a working car engine cannot be created by the car itself. Such a description can be created by someone who made the car. Therefore, the laws of nature existed before the Universe, as a kind of plan in some immaterial and timeless "medium" (we could not find a better word!). There should be information associated with the plan of how things should work in nature.

According to this concept, the laws of physics are not a subjective description of the world. They do require some numeral system, which could be binary, decimal, hexadecimal, and so on. Newton's second law, *"F=a × m"* (where *"F"* is the force applied to the body, *"a"* is the acceleration of the body, and *"m"* is the mass of the body), would look the same even to snails if they evolved to a level where they could discover this law. All that would change are the numeral symbols used. Perhaps they would have a binary numeric system, depending on the number of their antennas, unlike humans with a decimal system corresponding to the number of fingers. Maybe they would record the law more intricately, as a generalization of some other law. But the essence of the law itself would not change. Such laws are inherent properties of nature itself, and we merely discover them.

Here is what Einstein wrote to M. Berkowitz in 1950 (Hermanns 1983):

"God' is a mystery. But a comprehensible mystery. I have nothing but awe when I observe the laws of nature. There are no laws without a lawgiver, but how does this lawgiver look? Certainly not like a man magnified".

But here's what's amazing: Why is it that, in most cases, do these laws so seamlessly and perfectly align with the abstract mathematical constructs humans have devised?

Let's take the simplest example — Newton's law of universal gravitation. It simply states that the force of gravitation F is directly proportional to the product of the masses $M1 \times M2$ of the two bodies and inversely proportional to the square of the distance R^2 between them:

$$F = G \times \frac{M1 \times M2}{R^2}$$

where "*G*" is the gravitational constant. Why is this formula so simple, and not a very complex dependency? Or a relationship for which it is impossible to find any analytical expression at all? After all, one can find so many more complex dependencies leading to more or less similar observed results! But no, — nature "decided" that the simplest expression is just right. And here is an even more complex scenario: Imagine a world where the relationship between force and distance depends on time!

It can be imagined that any other form of the law of gravitation might lead to some instabilities in our solar system, and in the Universe as a whole. However, the question remains: Why are such simple expressions used to describe natural laws? After all, many natural laws could have been represented by infinite Taylor series[5] with many corrections, which would have halted progress for many centuries. There is a huge list of physical and chemical laws that look utterly simple and are expressed by simple mathematical functions, which were invented long before all these laws were discovered.

Looking at the history of science, one can easily conclude that logically consistent mathematical constructions, created out of principles of beauty, have turned out to be incredibly suitable for describing nature. Initially, such constructions arose in the mind, and even their creators considered these mathematical abstractions too speculative and having nothing to do with reality. However, these mathematical structures were too beautiful to ignore. And later, in some incredible

---

[5] The Taylor series is a decomposition of a function into an infinite sum of power functions.

way, we find that they are absolutely accurate for describing natural phenomena. There are countless such examples. I will just mention a few cases from the field of physics.

British theoretical physicist Paul Dirac (1902 – 1984) was one of those who relied heavily on mathematical logic and the beauty of abstract mathematical equations. Dirac simply assumed that he is on the right track to find something new if the equation is mathematically beautiful and simple. He was more of a mathematician than a physicist.

In 1928, Dirac proposed an equation to describe the electron. However, this equation had two solutions: One for the electron with positive energy and another for the electron with negative energy. But physics (and common sense) implied that the energy of a particle should always be a positive number. Dirac interpreted the solution with negative energy as the existence of antiparticles. They were exactly analogous to particles but had the opposite electric charge. In the case of the electron, there must exist an "antielectron", identical in all respects but with a positive electric charge. Such a particle was named a positron.

At that time, this idea seemed too far-fetched and unrealistic. Even Dirac himself initially couldn't believe that such a mathematical trick could be connected to reality in some way. However, in 1932, shortly after the prediction about positrons, American experimental physicist Carl Anderson (1905 – 1991) discovered such particles in cosmic ray collisions.

Here is another example. The Norwegian mathematician Marius Sophus Lie (1842 – 1899) achieved significant success in research that led to mathematical formalism now known as "Lie algebra" or "Lie groups". His work described the properties of infinitesimal rotations. However, in the 18th century, his articles were very abstract and not at all connected to the description of nature, so not many readers could understand what exactly was being discussed. Partly, this happened because he had a unique style of writing, and his geometric intuition in mathematics far surpassed the intuition of other scientists. Now, this formalism is present in the vocabulary of any physicist. It turned out that Lie groups are the most elegant way to unravel the reality of the quantum world of elementary particles.

Here is a more recent example, which has been discussed before. In 1964, physicist Peter Higgs (1929 – 2024), along with other scientists suggested that there exists a special field, interaction with which gives elementary particles their mass. Higgs wrote a paper describing his abstract theoretical model, but the paper was rejected by the journal as "not having obvious relevance to physics". Then, Higgs revised the paper and submitted it to another journal, where it was published. Later, this field was named the Higgs field, and the particle — the carrier of this field — was named the Higgs boson. This heavy particle remained a purely hypothetical mathematical construct for a very long time. Many physicists doubted Higgs's theory. Nonetheless, in July 2012, scientists working at the Large Hadron Collider announced the discovery of this particle. All its quantum properties matched the predictions precisely.

The examples above illustrate a remarkable phenomenon: The seamless integration of abstract constructs into descriptions of nature. After all, the human brain is capable of creating abstract mathematical models that, ultimately, describe the fabric of reality. As we've said, laws must precede nature. Nature cannot create laws, as it already operates according to such laws. The fact that we can derive such laws, following abstract logic and mathematical intuition, suggests that an intelligent design preceded nature. This also suggests that our mind had some relation to this plan before our environment emerged from the chaos of the Big Bang.

177

American theoretical physicist John Wheeler (1911 – 2008), who coined popular terms like "black hole" and "wormhole", proposed that information is a fundamental concept in physics and its laws. Reality itself is created by observers in the Universe. He believed that every object in the physical world, in most cases, has a very deep immaterial source and explanation (Wheeler 1990). Accordingly, laws are informational products, created before the birth of the Universe.

7.3 Again about the origin of life

As we discussed earlier, the problem of life is one of the most serious questions which remains unsolved. How did molecules begin to combine into the simplest cells, which are incredibly complex "factories" for storing and processing information? One can go even further and ask: How did a cell lead to complex microorganisms? Why would a cell evolve? We understand well that a cell is not sentient and cannot make any decisions. However, the mechanism of self-organization into huge colonies of cells was somehow predetermined. For better survival? But a cell is not sentient, so why would it care about survival? It has no fear of ceasing to exist. Or let's suppose, all non-viable conglomerates of cells simply disintegrated, leaving no offspring. But a conglomerate of clumped cells is not yet a microorganism. And how do we get such complex organisms as animals from elementary organisms? Where did all this additional information come from, which is necessary for creating such complex systems? We have already discussed this question in detail in chapter 3.6 Theory of evolution.

As we have already noted, the problem of the origin of life is the creation of new information from components of the material world that, on their own, are incapable of creating information. This problem exists only when we consider that there is an arrow of time, proceeding from the past to the future. There are causal relationships. When an event occurs, it influences the future. In the past, chemical substances assembled into individual molecules, cells and organisms, creating vast amounts of diverse information, and this has influenced how the world looks now. We do not know how this self-organization process occurred.

However, if we consider the events of the emergence of life without the arrow of time, that is, taking into account the impact of synchronizing events on both the past and the present without distinguishing between past and present, then the problem of the origin of life becomes more solvable. According to this hypothesis, the fact that a complex organism exists now and looks the way it does also affects the past and causes unusual coincidences of non-living matter to combine into living organisms for the existence of organisms in the present. The fact that evolution proceeds from simple to complex is explained by the future complex organisms influencing how their formation should occur in the past. This happens at moments when organisms face threats to their existence. Even simple animals have an awareness of their existence and strive to survive. Of course, this does not negate evolution itself. Simply put, moments of synchronicity lead to "bursts" of new information in the past and the "acceleration" of evolutionary processes so that they occur in the desired direction for the future of such organisms. These synchronization impulses may look like "explosions" of new information, as shown in Figure 3.6 of chapter 3.6 Theory of evolution.

Is such a hypothesis scientific and can it evolve into a theory in the future? We do not know. There are about 20 explanations (Hands 2016) for how complex organisms emerge from simpler ones. Translating them into the category of theories is highly improbable.

All of them will likely remain hypotheses for many decades. Remember, we are dealing with historical information-rich events. They are no longer present. We cannot perform measurements to verify them. Everything we can assume now based on our observations cannot serve as strict proof that events have unfolded in a particular way over billions of years.

8 Waves of quantum mechanics probabilities

8.1 Collapse of the wave function

From the very moment quantum mechanics was discovered at the beginning of the 20th century, it became clear that something was amiss with the description of this world. It was found that the state of particles in the microworld is described by wave functions, knowledge of which allows for the most complete information about the probability of particles being in a specific location in space and their evolution over time. Austrian theoretical physicist, one of the creators of quantum mechanics, Erwin Schrödinger (1887 – 1961) derived an

181

equation that now bears his name. This equation makes predictions for wave functions. Obtaining such a function, as a solution to the Schrödinger equation, allows for finding the probabilistic behavior of particles. The square of the modulus of the wave function determines the probability density that a particle can be found at a point in space with certain coordinates at a certain time. However, the wave function itself involves imaginary numbers. Generally speaking, such functions have little to do with descriptions of things we deal with in everyday life. See chapter <u>8.5 The reality of imaginary numbers</u>.

If particles are described by probability waves, this means that one can never be certain about both the position and the velocity of a microscopic particle at a given time. According to the uncertainty principle proposed by the German physicist Werner Heisenberg (1901 – 1976), the more precisely the position of a particle is known, the less precisely its velocity can be determined, and vice versa. This has nothing to do with the process of observation. This uncertainty is a fundamental property of quantum systems, whether we observe them or not, and a consequence of the wave nature of matter. In other words, a wave is not located at a single point in space.

Another aspect of this principle relates to the uncertainties in simultaneous measurements of the energy of a microscopic particle's state and its lifetime. The energy-time uncertainty relationship is a fundamental principle of quantum mechanics, stating that the more precisely the energy of a particle is known, the less precisely its lifetime can be determined, and vice versa. In other words, there is an inevitable compromise between the accuracy of energy and time measurements.

The wave properties of particles are incompatible with the notion of their movement along specific classical trajectories. This oddity in the behavior of microparticles has troubled researchers for many years. No one knows what a wave function truly is.

However, real measurements of particles always detect the physical system in a specific state. It is said that the process of registering a particle "collapses" or "reduces" the wave function. And the act of measurement itself is the collapse of the wave function. Thus, the collapse can be defined as the transition between the potential (described by a wave) and the actual property. When a measurement is conducted, particles are forced to "choose" a specific state. It's as if something (or someone!) external decides how to display results and what exactly we should see.

The primary interpretation adopted by most physicists is called the "Copenhagen Interpretation", proposed by Danish physicist Niels Bohr (1885 – 1962) and Werner Heisenberg. It posits that the wave-like probabilistic behavior of particles "collapses" upon observation. And the wave function is nothing more than an abstract mathematical concept that simply reflects our uncertainty and lack of knowledge before observation. There's no point in wondering about what happens in some unseen "quantum wave" involving complex numbers. Quantum mechanics itself is subjective, as the wave function represents the observer's knowledge about the quantum system, and the collapse of the wave function is a subjective update of probabilities when the observer acquires new data.

The only flaw of this idea is the lack of information about what happens during the collapse of the wave function. Quantum mechanics does not address this. Furthermore, it's unclear where the alternative states of the quantum system disappear. This conundrum has been dubbed the "measurement problem".

Another popular (materialistic) interpretation is called the "many-worlds interpretation", or "multiverse interpretation", proposed by American physicist Hugh Everett (1930 – 1982). The term "many-worlds" owes its existence to another American physicist, Bryce DeWitt (1923 – 2004), who developed the theme of Everett's original work. He postulated that every time a measurement is made, all possible outcomes occur in different branches of reality, creating a multitude of parallel universes. This can be imagined as the "splitting" of the observer into clones, each seeing different measurement outcomes. In our view, this is an overly complex theory that also fails to explain the mechanism of the reality branching processes and where all these universes exist. Do all these universes change for everyone, not just for the person observing the quantum process? In any case, explaining one unknown with an infinite number of unknown universes seems far-fetched. We have made this point before.

But perhaps, collapsing the wave function is a property exclusive to consciousness? This does not fit at all with the concept of objective reality and materialism. Materialism requires a theoretical construction that is not related to humans as thinking beings.

That's precisely why another interpretation, known as the von Neumann-Wigner interpretation, has been criticized. It suggests that the consciousness of the observer is necessary for the collapse of the wave function. A human observer is necessary not just for observing the properties of an object, but even for defining these properties, with consciousness posited as necessary for completing the quantum measurement process. Consciousness serves as a mere bridge between materialism and idealism. Nonetheless, many renowned scientists support this interpretation.

For instance, British theoretical physicist Roger Penrose is among those who advocate for the inclusion of the "phenomenon of consciousness" at the foundation of quantum mechanics theory. In his book "The Emperor's New Mind: Concerning Computers, Minds, and the Laws of Physics" (Penrose 1989), he agrees with the idea that mathematics is not a construct of our consciousness but a manifestation of a world of mathematical ideals. The scientist leans towards the notion that the physical world is a manifestation of a timeless mathematical world of ideas. See chapter 3.9 Idealism and information. Consequently, human mental activity cannot be fully described by computer algorithms. As the reader may have already understood, I also lean towards this concept of quantum mechanics. I think that most of the observations presented in this book are consistent with this view of the world.

Unfortunately, for practical purposes, all these interpretations of quantum mechanics are indistinguishable. They all assume the same outcomes for observations in quantum mechanical experiments.

8.2 Nonlocality

In our world, a signal cannot propagate faster than the speed of light. It corresponds to the speed at which electromagnetic waves propagate in a vacuum. The speed of light, usually denoted with the letter "c", is approximately equal to 3×10^8 meters per second (m/s). Therefore, any influence on a distant part of the world can only have an effect after the time it takes for a signal to travel from one point to another at the speed of "c".

In the quantum world, the situation is different: Influences can act instantaneously over arbitrarily large distances. In quantum physics, this is referred to as "nonlocality" or "quantum entanglement". This phenomenon occurs when the states of several particles become interrelated regardless of the distance between them. For example, when two interacting particles are separated from each other, they can remain "entangled". They behave as if they were one entity, no matter how far apart they have moved from each other. In physics, this means that measuring a property of one of the particles instantly sets the value for the other, no matter which part of the Universe it has moved to.

This creates a dilemma: The possibility of faster-than-light communication. One explanation for this effect is to hypothesize that particles have hidden variables that describe them with absolute precision. This information about the particles cannot be revealed because it does not belong to the physical Universe. You've probably played table tennis ("pingpong") on a computer, where the ball flies from one edge of the screen to the other. The ball indeed travels distances measured in hundreds of pixels and tens of centimeters on the screen surface. In reality, its description is located in one place, somewhere in a tiny spot on the chip. All its movement is determined by an algorithm in the microprocessor, for which screen distances are irrelevant. This algorithm knows everything about the movement of this ball, but this information is inaccessible to you if you are just observing the computer screen.

One can also look at the "quantum entanglement" problem differently. Imagine that causality could go in the reverse direction. This would mean that a particle could transfer the action of its measurement back in time, to the moment when it was entangled, affecting its partner. No messages traveling faster than the speed of light are required. Instead of having non-local connections between particles separated by infinite distances, perhaps the connection goes through time, from present into the past.

Indeed, some interpretations of quantum mechanics suggest that the act of observing or measuring a particle can affect the information obtained by the observer about the past. This is called "retro-causality". There are even some experiments that show quantum entanglement can occur through time, not just space. This challenges our understanding of causality, which posits that an effect cannot occur before its cause. If quantum entanglement can occur through time, it might mean that the present can affect the past.

Stephen Hawking was convinced of the retrocausality of the Universe. He stated:

"Quantum physics tells us that no matter how thorough our observation of the present, the (unobserved) past, like the future, is indefinite and exists only as a spectrum of possibilities. The universe, according to quantum physics, has no single past, or history. The fact that the past takes no definite form means that observations you make on a system in the present affect its past" (Hawking and Mlodinow 2010).

Let us recall that the laws of physics prohibit time travel to the past for many reasons. If we were to travel back in time and alter the course of events, we would change the course of history. There is no evidence that we are currently inundated with any such future messages or messengers.

The "retrocausality" hypothesis does not assume that signals or objects, including humans, can be sent to the past. Retrocausality is not time travel. It's a mechanism that allows future circumstances to correlate with past states. This means that if you are conducting an experiment and obtain a certain measurement, all circumstances of this measurement must have a past. But the information about the past can change in such a way that you obtain exactly the measurement you are observing in your present. In other words, when the experimenter chooses the parameters of the equipment for measuring particles, this decision affects the properties of these particles in the past. The human choice made in the present can influence something in the past. It doesn't mean that the past has changed. The past doesn't exist in a way that we can observe or measure it. All that remains is information about the past, which is "synchronized" with the observer from the present.

All of this might suggest that the microworld is somehow linked to a reality where neither time nor space exists. That reality contains all the information about the particles of this world. We only observe some consequences of what happens in that other reality, where there is no arrow of time. We can only infer some incomplete knowledge about particle movements, while the real "mechanism" that governs the motion of particles is hidden in the information reality that has the full description about dynamics of particles.

This is a complex and still intensely debated research area in physics, so we won't delve further into it.

8.3 Non-material forms of information

Carl Jung believed that our minds are guided by a system of conceptual forms or archetypes — elements of the collective unconscious. These are real, albeit invisible or measurable by any instrument. Quantum phenomena and the very basis of the material world are not material. They are determined by immaterial forms, without mass and energy. These forms have the potential to appear in the empirical world and act upon us. They create a world of "potentialities" of physical reality, and all empirical things are merely appearances of these invisible forms. These abstract forms are patterns of meaningful information, akin to thoughts. We've mentioned that this concept resonates with the idealism of Plato and many other thinkers (see chapter 3.9 Idealism and information).

As it turned out, Carl Jung's revolutionary views on the human mind may be consistent with the discoveries of quantum physics (Ponte and Schäfer 2013). Elementary particles, such as photons and electrons, are described by probability waves or probability fields. Probabilities are dimensionless numbers that carry no mass or energy, only information about numerical relationships. Nonetheless, the visible order of the world is determined by the interference of these waves. Thus, one may conclude that the "basis of reality" is immaterial.

An interesting discussion of this issue can be found in the book "Infinite Potential" (Schäfer 2013). It proposes that some "entities" lay the foundation of the visible world. They are pure forms in the guise of patterns of semantic information. They are not material and can be described by waves carrying information, which are immaterial and do not have a specific location in space but have many potential positions. These wave configurations are in a state of potentiality and are not part of the empirical world. When we interact with them, they appear to us as material "elementary things". Thus, particles obtain certain masses at specific points in space. After we stop interacting with them, they transition to a wave state and leave the empirical world. The Universe itself is a background of immaterial forms, not things. Forms are real, although invisible, as they have the potential to appear in the empirical world and act within it. This approach aligns well with the von Neumann-Wigner interpretation and the further developed theory of the "phenomenon of consciousness and mathematical forms existing outside of humans" (Penrose 2007).

8.4 Problems of interpretations

As we said earlier, all interpretations of quantum mechanics are empirically indistinguishable because they all predict the same outcomes for quantum mechanical experiments. This means that the problem of explaining quantum mechanics is not entirely scientific. It depends on a system of beliefs and philosophical perspectives.

Which interpretation of quantum mechanics is more likely? Maybe the clue can be found not in empirical observations but in the analysis of the formulation of quantum mechanics itself and how it was discovered? Maybe our understanding of the amazing coincidences in the laws of nature will allow us to solve the problem of the interpretation of quantum mechanics?

Imagine this situation: In some country, a long time ago, say, in the 16th century, some eccentric came up with a fantastic world of the future. This fictional world had no connection to that country or to the 16th century. In this imagined world, there are spaceships and magical kingdoms. Even the language spoken by these people of the future is completely different from the language spoken in that time. This eccentric individual, with strong determination, develops his fantastical world, following principles of logic and beauty. Everyone laughs at him and urges him to write a book about the real life around him, which he fails to pay attention to. But no — he stubbornly writes about his fantastic world, which has nothing in common with reality and exists only in his imagination. Moreover, he finds followers, who even help improve and add details to this imagined world.

Centuries passed. Humanity embarked on star travels. But strangely, the first civilization they encountered looked exactly as described by that strange person from the 16th century. The same names of continents, the same names of rulers, and the same incredible machines. And even a similar language!

What a strange coincidence!? One natural explanation could be that the 16th-century person was not a fantasist. He simply analyzed how a future civilization might look and how language might evolve.

Here, we must say — it is indeed possible to arrive at some logical conclusion about what machines might look like in the future. Yet, it's far from evident how one could derive the names of continents and even the language itself from notions of logic and beauty.

Another possible explanation is this: Somehow, information from that civilization made its way to Earth in the 16th century, and somehow, this person was able to understand it and describe it in his stories. But we know for certain that there was no exchange of information through interstellar messages. This means the only possibility is that the information was transmitted by some other means. Perhaps it came from a common pool of Universal information to which the mind could "tune in".

Let's return to quantum mechanics. Ask yourself this question: Why does quantum mechanics use complex numbers to describe our world? After all, these numbers were invented without any practical purpose long before the discovery of quantum mechanics. We will talk about this in the next chapter.

8.5 The reality of imaginary numbers

The strangest thing about quantum mechanics is that its mathematical description is based on "imaginary" (or complex) numbers, which become negative when multiplied by themselves. These numbers didn't make any sense at the time of their invention in the 16th century. It should be noted that the equations of quantum mechanics can be rewritten using real numbers, but this would be so inconvenient and cumbersome that such a formulation would not become widely adopted.

The birth of the complex number was historically linked to the desire to "legalize" the square roots of negative numbers in the 16th century. However, it was only in the 1600s that French mathematician René Descartes (1596 – 1650) developed his rule for such numbers and mentioned the term "imaginary" — imaginaires ("imaginary" — a more common usage nowadays) for complex numbers. Subsequently, a famous German mathematician Gottfried Leibniz (1646 – 1716) provided a compelling reason why these numbers are "imaginary" (Kline 1972). He noted,

"The Divine Spirit found a sublime escape in the miracle of analysis... that amphibian between being and non-being, which we call the 'imaginary root of negative unity'."

Subsequently, a new stimulus for the study of complex numbers as a separate subject emerged in the 16th to 18th centuries, when algebraic solutions for the roots of cubic and quartic polynomials were found. However, their utility for describing the observable world was not understood. Leonhard Euler (1707 – 1783), a Swiss, Prussian, and Russian mathematician, wrote: *"For we can assert that they are nothing, no more than nothing, and no less than nothing, which necessarily makes them imaginary or impossible"*. The French mathematician Augustin-Louis Cauchy (1789 – 1857) said that he "dismissed" the imaginary unit *"without regret, because we do not know what this supposed symbolism means and what significance to attribute to it"*.

Complex numbers are written as: "$a + b\,i$" where "a" and "b" are real numbers, and "i" is the imaginary unit, i.e., the number for which the equation

$$i^2 = -1$$

holds true. Just like for regular numbers, operations of addition, subtraction, multiplication and division are also defined for complex numbers.

The emergence of complex number theory among mathematicians between the 17th and 19th centuries is a somewhat unexpected development. People started studying these numbers guided solely by notions of mathematical beauty. Of course, one could argue that all the logic of mathematical thinking led to the idea that complex numbers were useful. But that's not entirely the case.

Almost all mathematical constructs of that time had a quite specific meaning tied to reality. For instance, negative numbers are mentioned as early as 200 BCE in China, where they were used to represent the amount of debt. It's simply the quantity of what you do not have, or what you are lacking. It is assumed that in the 7th century CE, negative numbers were described by the Indian mathematician Brahmagupta. But even earlier, they were mentioned by the ancient Greek mathematician Diophantus (3rd century CE).

There was also nothing surprising about irrational numbers. We recall that an irrational number is a real number that cannot be written as a simple fraction. The earliest known use of such numbers appears in Indian texts, written around 750 BCE. For ritual sacrifices, it was required to construct a square fire altar whose area was twice that of a given square altar, leading to the finding of the value of $\sqrt{2}$. In literature, it was named Pythagoras' number (Agarwal and Agarwal 2021).

The development of the formalism of imaginary numbers began much earlier than the creation of absolutely necessary mathematical concepts describing our world such as differential calculus, ordinary differential equations (1629), probability theory (1654), Taylor series (1712) and the normal distribution (1733). At that time, many considered "imaginary" numbers to be fictitious and useless.

But everything changed with the advent of quantum mechanics. From its very inception, it operated with complex numbers. Eugene Wigner (1902 – 1995), a Hungarian-American theoretical physicist, who won the Nobel Prize in Physics for his contribution in quantum mechanics, wrote:

"The use of complex numbers in quantum mechanics is not a mere computational trick of applied mathematics; they enter into the very essence of the formulation of the fundamental laws of quantum mechanics".

In electromagnetism (the study of electromagnetic phenomena) and in most other areas of physics, imaginary numbers are simply a mathematical convenience. But not in quantum mechanics! As I've already mentioned, from the very beginning, pioneers of quantum mechanics refused to attempt to develop a quantum theory based on real numbers. They believed that ordinary (real) numbers were impractical for the quantum world. However, the possibility of using real numbers was never formally excluded. Quite recently, two independent studies showed that to reproduce experimental results, a formulation of quantum mechanics in complex numbers, not real numbers, is necessary (Miller 2022) (Avella 2022). As Penrose noted earlier, complex numbers are more fundamental than real numbers. Perhaps they even precede real numbers (Penrose 2007).

Now ask yourself — was the discovery of complex numbers a coincidental event for the formulation of quantum mechanics — a physical theory that operates with imaginary waves, is described by imaginary numbers invented without any practical reason, and which can be interpreted as necessitating the existence of consciousness?

For many, the answer might be obvious. Purely imaginary, illusory numbers, having no relation to the reality with which we deal on a daily basis, turned out to be the natural language for describing the microworld. Such a coincidence might suggest the necessity of consciousness in the interpretation of quantum mechanics. Our consciousness has access to a certain reality where complex numbers are the natural language on which quantum mechanics is built upon. Perhaps our minds find comfort and beauty in such numbers simply because our consciousness is still connected to the original source where these numbers were first used. And once we discovered quantum mechanics and were able to perform quantum measurements, consciousness becomes a necessary component that translates illusory wave functions into real observable quantities.

This is the von Neumann-Wigner interpretation, which suggests that the collapse of the wave function requires the consciousness of an observer. Therefore, it's not surprising that consciousness possesses the ability for super-intuition, capable of inventing mathematical descriptions and seeing something important in them, even though it makes no sense for ordinary things we are dealing with in this reality. Many may argue: The quantum world and consciousness are interconnected.

Therefore, it's no coincidence that theoretical physicist R. Penrose, along with anesthesiologist S. Hameroff, proposed that the human mind has a quantum nature and described a possible quantum mechanism in the human brain. Penrose had previously suggested that consciousness is not entirely algorithmic — it's integrated into the quantum world (Penrose 1989). We won't delve into the quantum mechanisms in the brain proposed by Penrose and Hameroff. It's an interesting hypothesis that is discussed in the book (Mensky 2011). Consciousness explained as a purely quantum phenomenon was also discussed by Federico Faggin, Italian-American physicist, known for designing the first commercial microprocessor. According to his ideas, the physical world is a virtual reality system constructed by "selves" (Faggin 2024).

If this hypothesis is true, and our consciousness is somehow connected to quantum-mechanical effects, it could potentially explain synchronicity. We've already discussed that synchronicity might be related to the modification or editing of information about the past (see chapter 6.9 Nothing is accidental). This very property could be used to explain "nonlocality" or "entanglement" of elementary particles in quantum mechanics (see chapter 8.2 Nonlocality). This suggests a profound connection between the phenomena of quantum mechanics and the operations of consciousness, potentially bridging subjective experiences with the objective realities of the physical world.

One might object that quantum effects manifest on very small scales, while our brain is composed of cells ranging from 1 to 100 micrometers in size. However, if our brain acts as a device that has access to the quantum world, thanks to some collective effect of the order of a hundred billion neurons, then such a hypothesis is not entirely fantastical. As we recall, in the "many-worlds" interpretation of quantum mechanics, every time a measurement is made, all possible outcomes occur in different branches of reality, creating a multitude of parallel universes. This hypothesis appears even more fantastical than the theory suggesting our brain has direct access to the reality of abstract forms, using quantum-mechanical effects. Interaction with this reality creates the effect of consciousness, which is capable of "editing" the past in moments of crisis to achieve the best possible outcome in the present. Maybe retrocausality plays a role here, as we discussed in chapter 8.2 Nonlocality. The same hypothesis explains why the observer and their consciousness are so crucial in interpretations of quantum mechanics.

8.6 Again about information

Well known American physicist John Wheeler (1911 – 2008) suggested that information is a fundamental concept in the microworld (Wheeler 1990). Every equation of quantum mechanics contains a physical quantity called Planck's constant. This constant is also referred to as the elementary quantum of action. Recall that in the microworld, energy is transferred in discrete packets – quanta. Their energy is astronomically small. The energy carried by a single quantum is $E = h \times v$, where "v" is the frequency of radiation (wave), and "h" is the elementary quantum of action. Thus, "h" links the quantum of energy of a quantum system to its frequency. Planck's constant is approximately equal to:

$$h = 6.626 \times 10^{-34} \text{ joule-seconds (J·s)}$$

assuming the International System of Units.

For a person, it is very difficult to imagine such a small value. For example, the inverse diameter of the observable Universe is about $1/(94 \text{ billion light years})$, which is "only" about 10^{-27} m^{-1}. Here the unit "m^{-1}" denotes the inverse meter.

It can be said that "h" defines the lowest limit of spatial magnitude, starting from which the world of the microcosm comes into effect. It determines the boundary between the macrocosm, where Newton's laws of mechanics operate, and the microcosm, where the laws of quantum mechanics with its Heisenberg uncertainty principle are in effect (see chapter <u>8.1 Collapse of the wave function</u>).

According to Wheeler, every time we see an equation with this constant, it's a sign of the presence of the smallest quantum of information. As we remember, a "bit" is the minimum basic unit of information measurement. Wheeler said,

"It's from a bit... Formulas with this magnitude reflect a part of physics that we have learned to translate in terms of information theory. Tomorrow we will learn to understand and express all of physics in the language of information".

He suggested that we have reached the very foundation of the world, since the laws we derived contain this constant. All objects in the Universe should be considered as secondary, created from carriers of an abstract and fundamental entity — information. The deepest foundation of the physical world is information itself. It does not come from matter.

Many contemporary physicists share a similar viewpoint. For example, the British physicist and mathematician Stephen Wolfram firmly believes that all features of our Universe actually arise from an ultimate discrete (discontinuous) quantity that must be astronomically small (Wolfram 2002).

9 Coincidences in numbers

     Coincidences without an apparent reason to appear are the first sign of a hidden mechanism that may cause such phenomena. This is also true for logically related numbers used by humans to describe nature. In chapter 3.10 Remarkable number, I talked about the number 6, which appears in all places associated with the emergence of life. Further, in chapter 4.3 The birthday problem, I explained that one should be cautious regarding coincidences in numbers. It's easy to find amazing correlation patterns purely from the law of large numbers. In reality, true coincidences are those that have connections with important logically-connected categories. This was discussed in chapter 6 Examples of significant coincidences.

In mathematics, there are quite a few universal numbers that are exceptionally frequently used across all fields of science. Arguably, the most commonly encountered numbers are:

- The number "pi" (denoted by the Greek letter "π"). Its value is approximately 3.14159. This constant corresponds to the area of a unit circle, the half-period of trigonometric functions, and much more.
- The number "e" — the base of the natural logarithm. Its value is approximately 2.71828. This constant is sometimes called Euler's number.

These numbers are tabulated in mathematical programs and reference books. There are a few other constants that are encountered fairly often, but they are not popular enough for us to discuss here. We will also not cover constants that are used in certain areas of physics, chemistry and other natural sciences.

Since there are relatively few such fundamental constants used by humans, any coincidences involving these numbers must be quite interesting for our discussion. Here we are not dealing with a large static sample (or possibilities) for calculating probabilities. Our set of numbers is very small, so unlikely matches, correlations between them or unusual features should attract close attention.

These constants are irrational. This means each of them consists of an infinite series of random numbers. An irrational number is a real number that is not rational, that is, it cannot be expressed as an ordinary fraction A/B.

For the laws of nature, there is absolutely no need to have so many digits after the decimal point for the most fundamental constants. Nothing in the world would change if these digits were "shortened". However, since they are infinitely long, perhaps they might serve as a "container" for some information. If there is an intelligent design in nature, and the designer decided to reveal themselves by leaving us a message, the presence of any regularity in such numbers may be the key to understanding the design of this world. Why not? In this book we ask questions that go beyond science.

9.1 The number π

Certainly, almost everyone is familiar with the number "π". This mathematical constant represents the ratio of a circle's circumference to its diameter. It is approximately equal to 3.14159265358979323846264... (and so on). The first 100 trillion digits of "π" after the decimal point are already known. As we said before, the number "π" is irrational, meaning its value cannot be precisely expressed as a fraction, and its decimal representation never ends and is not periodic.

The discoverers of the number "π" are believed to be people from prehistoric times. They noticed that to make a basket of the desired size, one had to take rods three times longer than its diameter. The ancient Babylonians knew about the existence of the number "π" nearly 4,000 years ago. The number "π" is even found in the dimensions of the Great Pyramid at Giza. As it turns out, it has the same ratio of height to the perimeter of its base as the radius of a circle to its circumference, that is, $1/2 \times \pi$. The number "π" has no dimension, making it difficult to meaningfully relate it to any law in physics or chemistry.

The number "π" is considered "normal". This means that its digits are random in a certain statistical sense. However, there is one interesting feature: This number has a sequence of six nines (999999), starting from the 762nd decimal place. Sometimes this feature is named the "Feynman Point" — after the famous American physicist Richard Feynman (1918 – 1988) . However, it is possible that the earliest mention of this feature of the number "π" in literature can be found in Douglas Hofstadter's book "Metamagical Themas" (Hofstadter 1985).

The most remarkable fact of this observation is that the position of the number "999999" appears very early in the sequence of random numbers. For a normal number, chosen uniformly and randomly, the probability that a specific sequence of six digits will appear so early in the beginning of the decimal representation is about 0.0008 (Arndt and Haenel 2001). We also calculated such a probability using a computer code and arrived at the same conclusion in chapter 20.9 Appendix.

It is interesting that all other numbers with six repeating digits occur much, much later. The appearance of such a number so early can be explained by the enormous variation of the distribution of the so-called expected positions of groups with the same digits. Let's demonstrate the first positions of occurrence for repeating combinations of six digits:

- **111111** at position 255,945
- **222222** at position 963,024
- **333333** at position 710,100
- **444444** at position 828,499
- **555555** at position 244,453
- **666666** at position 252,499
- **777777** at position 399,579

- **888888** at position 222,299
- **999999** at position 762 (!)

As we can see, indeed, the number "999999" appears much earlier than all other repeating digits. A naturalistic explanation for this is pure randomness.

It's interesting to note that other known constants, such as the golden ratio (approximately equal to 1.618), do not have the number "999999" so close to the beginning. In the case of the golden ratio, this number is at position 1,955,975. But the golden ratio is much less frequently used.

We should, however, note that there is a large number of six-digit combinations that have a smaller probability of appearing earlier than what we expect for natural random numbers. In chapter 20.9 Appendix, we find all the six-digit numbers that are even more unlikely to appear earlier than 999999. The number of such numbers is about 500. Their appearance of such numbers at a position smaller than or equal to 762 has a probability of less than 0.0008 (in relation to random appearances of digits).

For instance, the number 589793 occurs at position 12 with an incredibly small probability, about 9×10^{-6}. It is also a remarkable number that stands out from all other random and irrational numbers, as it is rarely encountered. It might seem that the question can be closed. For science, the position 762 of the number 999999 is a pure coincidence. There are other more improbable numbers happening at earlier positions. The pure naturalistic explanation for the six nines is as follows: Six nines appear at the 762nd digit of the value of "π". That's all. A mere artifact of the decimal system. If one were to write "π" in hexadecimal format, there may be other coincidences.

But it's not that simple. This book is not scientific, and thus we can pose questions impossible for science to address. So far, we have assumed that coincidences are truly significant only in cases where there is a logical connection with something important. Is there a universally accepted significance to the number "999999"?

The number "999999" is not just some integer value. It is a number that must easily be found visually by humans when examining a random sequence of numbers. It is precisely such numbers that people will "hunt" for if they decide that an infinite sequence of numbers may contain some information, or message. Numbers like these serve as "markers" for those who choose to encode information. Let me explain why.

Among the 500 six-digit numbers that appear more rarely at positions less than or equal to 762, there is not a single number that is as simple and appealing for humans as "999999". All such numbers are complex and contain a random assortment of meaningless digits. It's notable that the number "999999" isn't as visually captivating in other numbering systems:

- In hexadecimal, it's f423f,
- In binary, it's 11110100001000111111.

Therefore, markers in other systems would be in different places, if people in the past could associate such markers with certain concepts. However, I haven't heard about such markers in other systems. This might not be a coincidence, as such systems have only recently come into active use in computers. They are not at all related to the ancient history of civilizations or the structure of the human body. Yet, if such a message is for people, a connection with humanity must exist somewhere.

Conversely, the decimal system was based on counting with ten fingers. From prehistoric times, such a system has been closely linked to humans and the structure of their hands. This number system can be traced back to Egyptian, Greek, Hebrew, Chinese and many other civilizations.

But let's go back to the decimal system. What's also interesting is this: The number "999" is also at the position 762. But the probability of finding it among irrational numbers at a position less than 762 is very high — 0.56. Modify the code in chapter 20.9 Appendix to prove this. Therefore, such a "label" is clearly not suitable. But a label consisting of two numbers of 999 will definitely work.

Try to use your imagination. Suppose the Creator (a higher intelligence?) designs a world and adjusts the values of all the constants on which all natural laws will function. They need some kind of analogue to a number that will link together certain quantities, such as the circumference of a circle and its diameter. This number is required to a certain degree of precision. Suppose the precision to the 172nd digit is sufficient for the stability of all parts and phenomena in our Universe. Note that even NASA (National Aeronautics and Space Administration) uses the number "π" with precision up to 16 digits after the decimal point for navigating satellites and spacecraft. Why not then place a "marker" in the number "π" in the form of an easily recognizable pattern of digits and encrypt some information using all the other numbers after this marker? Obviously, this shouldn't pose any issues for the laws of the Universe. And then do the same in other non-decimal numeral systems (of which there are not so many). However, it is precisely the decimal system that is ideal for such purposes, as it is based on the biological characteristics of humans. They, according to the plan, are supposed to solve this puzzle.

So, this mark must be highly recognizable and simple, ensuring easy identification by humanity. Precisely "999999" is an excellent candidate for such a purpose. There are 6 nines here. We have already discussed in chapter 3.10 Remarkable number, that this number is exceptionally important. It appeared in chapter 6.6 The Kaiser and the War. Miraculously, the number "6", in some mysterious way, connects the number "π" with the force of interaction between matter and light (see chapter 9.7 The fine structure of the world).

And so, we have dealt with the number "6". Then why does the number "9" repeat 6 times? The number "9" is a sacred number in Scandinavian and ancient Germanic symbolism. For Buddhists, nine is the highest spiritual power. In the Bible, it is the number of signs (miracles) given by the Lord to Moses (17:101; 27:8 – 12). Thus, six nines are a sign given to mankind.

The number "999" also has significance. This number has been imbued with meaning since time immemorial. "999" is the opposite of "666", and therefore it is often interpreted as a sign of God or spirit. In numerology, "999" is one of the strongest angel numbers, representing the completion of a cycle, the beginning of a new one, and divine guidance. The number "999999" carries the same meaning. There are very few numbers that are as significant as the number of nines.

Therefore, the group of six 9s is one of the most easily recognizable in the infinite series of random numbers of "π". Thus, the emergence of a "mark" that is statistically noticeable (compared to random numbers and irrational numbers) exists only in the decimal system, tied to human biology, and the presence of a meaningful concept associated with the numbers "6" and "9", may indicate that such a coincidence is not fully random.

Probably, few know that the number "π" appears in various unexpected contexts of descriptions of the natural world. An incredible number of formulas and equations of electronics, physics, and chemistry contain the constant "π". As we will see further, the number "π" appears in the descriptions of an infinitely large number of random events (chapter 9.3 Could this be accidental?), connects real numbers with imaginary ones (chapter 9.5 Everything comes together), and as a result — with the quantum world. With the help of the human-number - "6", the magnitude of "π" is connected with the fine-structure constant, which describes the force of interaction between light and matter (chapter 9.7 The fine structure of the world).

9.2 The number e

Now we are going to discuss the second most popular fundamental constant. As we said before, the constant "e" is of great importance in the mathematical description of the world. It is defined as the base of the natural exponential function. The first mention and approximate value of the number "e" dates back to 1614 in the works of the Scottish mathematician John Napier (1550 – 1617). The famous scientist Leonard Euler (1707 – 1783), one of the most prolific mathematicians of all time, used the symbol "e" in the theory of logarithms in 1727. Since then, the value of this constant has been continuously refined. The approximate value of the number "e" is equal to:

$$2.71828182845904 \ldots$$

Similar to "π", this constant is a "normal" number, meaning that the probability of different digits appearing in its representation is the same. This number is widely used in exponential functions, differential equations, in economics, in analysis involving complex numbers (more on this later), in probability theory, in the description of disease spread or radioactive decay and many other fields of human knowledge.

The constant "*e*" can be obtained by calculating the infinite limit of this expression:

$$e = \lim_{n \to \infty} \left(1 + \frac{1}{n}\right)^n$$

when "*n*" approaches infinity, denoted as "∞". Also, the number "*e*" can be obtained by summing a series of numbers:

$$\frac{1}{0!} + \frac{1}{1!} + \frac{1}{2!} + \frac{1}{3!} + \dots = e$$

to infinity. Here, the exclamation mark denotes the factorial. It's the product of all natural numbers from 1 to *n* (for example, 3! = 1×2×3 = 6). This expression can be conveniently represented as:

$$\sum_{n=0}^{\infty} \frac{1}{n!} = e$$

As we see, the value "*e*" can easily be obtained from an infinite series of numbers. In everyday life, people can't imagine what it means to deal with infinity. How to imagine an infinite number of things or universes (for those who like hypotheses with an infinite number of universes)? In religions, infinity is usually associated with God. In the case of the mathematical expression shown above, something incredible happened: We performed a mathematical operation with an infinite sequence of ordinary fractions, and we got a perfectly finite number — "*e*". The ability of the human mind to deal with infinities in abstract mathematical constructions, and obtaining a finite result, is one of the most amazing properties of thinking.

As you've already understood from the previous chapter, we're searching for patterns in irrational numbers. The larger the number of digits in the repeating segments of such numbers, the lower the probability that such a property will recur in a large set of random irrational numbers. Therefore, the uniqueness of such repeating numbers will increase. A simple analysis of this constant reveals that it contains two blocks of identical numbers: 1828. Indeed, pay attention to the part highlighted in bold:

$$2.7\textbf{1828}\textbf{1828}45904 \dots$$

For irrational numbers, the appearance of such a sequence seems quite suspicious, as such repetition occurred too early.

Let's calculate the probability of such a sequence appearing in many irrational numbers. In 20.10 Appendix, we provide the code for a program that calculates the occurrence of a block of any four digits at a position less than or equal to the position of occurrence in the number "e". This probability is approximately:

$$0.0003.$$

This is a sufficiently small probability, given that the number of invented combinations is not that large. Mathematicians understand that the repetition of a block of any four numbers in a random sequence of digits is unusual. The probability of such regularity appearing at the very beginning of a chaotic sequence of digits is one in several thousands. We numerically verified this in chapter 20.10 Appendix. Nevertheless, this phenomenon is often considered a common coincidence (Lange 2010). The appearance of such a block of numbers happened because something similar can always be found in random numbers if you look hard enough. That's the whole explanation.

With our code, we can even check where two identical blocks of 4 numbers appear in the number "π". Just put the number "π" in the code of chapter 20.10 Appendix. It turns out, the first such block (number 9314) appears at position 8238. And the probability of such an occurrence is 0.51. That is, not a remarkable observation for this number.

I want to point out one detail right away. I have no idea what's special about the number 1828. If you fantasize and assume that this is referring to the year 1828 CE, you will immediately discover that there were many different events that year. Even for the most enthusiastic numerologist, all historic events of the year 1828 are not so important to be marked in such a strange way. The only thing that arouses interest is that the two blocks with 1828 end at position nine. As noted before, the constant "π" has the number "9" in the group of six numbers that appear too early in the random sequence of digits. But maybe there's no connection here.

9.3 Could this be accidental?

Now we can calculate the probability that we have found blocks of repeating digits in the two most significant numbers used by people. This probability is equal to:

$$0.0008 \times 0.0003 = 2.4 \times 10^{-7}.$$

Note that we did not require specific numbers in the repeated blocks of digits. Of course, this opens up several new possibilities for constructing such blocks. For example, one could consider groups of 5 numbers (instead of four), as in the case with the number "*e*". But there will be very few such possibilities. Therefore, this will hardly change the obtained probability.

And now let's fantasize a bit and think about what this might mean. Suppose some cosmic intelligence decided to create the Universe and humans. But to do this, it had to create at least two universal numbers, on which the entire description of the world is built upon. These numbers are "π" and "*e*". They must be discovered by humans, who are created in the image and likeness of this intelligence. This means it's necessary to use the decimal system of numeration, as it most closely matches the biological structure of humans. This is the most natural, as such a system is likely to emerge when using ten fingers.

So, from several million possible irrational random numbers, this cosmic intelligence selects two numbers that have certain features. Namely, there are digital markers from two blocks of the number "1828" and two blocks of the number "999", which appear in the infinite sequence of these numbers too early compared to any other blocks of numbers.

But why should the marker blocks be so long? This is not too surprising. This is the most optimal so that such blocks can be found in an infinite series of random numbers. If we use two or three numbers, such regularities would appear too frequently, making it quite difficult to associate them with a specific feature in an infinite series of digits.

So, we've figured this out. The two most important numbers of human civilization, "π" and "*e*", contain markers that will only be understandable to humans:

999 999 and 1828 1828.

What do these numbers mean in such markers? As has already been said, the number nine is usually associated with the new divine beginning and the moment of something is finally completed. In numerology, "999" represents the pure form of a spirit, endowing mortals with the ability to comprehend the higher laws of the Universe. The repetition of this number was possibly necessary to notice this marker or to associate it with masculine and feminine principles. The number "9" also has significant meaning in Indian numerology, where it is believed to symbolize completion, achievement, and realization. It is on the ninth position that the last number "8" is found, which ends the block of 8 numbers (1828 1828).

Perhaps here I should draw a conclusion, as my imagination has almost run out. I cannot say that I have fully deciphered these combinations of digits. The assumption that they could be markers for decoding the code found in the infinite series of numbers of these amazing constants is not the craziest idea. At least for those who seek meaning in everything.

9.4 Infinity, sphere and randomness

What do you think will happen if we sum this series to infinity:

$$\frac{1}{1} + \frac{1}{2} + \frac{1}{3} + ... =$$

It turns out, you'll get an infinitely large number. We do not understand what infinity "∞" is. Infinity is impossible to comprehend and represent in our minds. However, abstract thinking can easily "compress" it into a finite number. We demonstrated how to do this in chapter 9.2 The number e using the factorial to get the number "e". And now, we will do it again by squaring each term. This is exactly what Leonard Euler did in 1735, and he got:

212

$$\frac{1}{1^2} + \frac{1}{2^2} + \frac{1}{3^2} + ... = \frac{\pi^2}{6}$$

Stop. What do we see? We got a finite number! But not just any number. Remarkably, we have obtained "π squared". What does "π", used to describe circles and sphericity, have to do with some infinite series of numbers? And then, surprisingly, the "number of man", 6, snuck into this expression. The same number that defines the size of the marker inside the never-ending mathematical constant "π". It seems, without "6", nothing at all is possible (see chapter <u>3.10 Remarkable number</u>).

It's difficult to call this a mere coincidence. The numbers "π", "*e*" and infinite series are somehow related. Mathematicians capture this connection in various expressions. Perhaps, the real meaning is hidden in the depths of formalism describing the structure of our Universe. For now, it remains beyond our understanding.

But let's go further. Have you ever thought about the connection between a sphere (or circle), infinity and randomness? These are three incredibly important concepts that must have been utilized by whoever planned and created this world, if such an assumption is true.

It turns out that there is indeed something common between these three abstract concepts. Again, it's the numbers "π" and "*e*"; two constants that are the most fundamental for our civilization. They also have markers from repeating groups of numbers, distinguishing them from many irrational numbers.

Let's repeat what we learned in the previous chapters.

Without the constant "π", it is impossible to work with circles, spheres and any spherically shaped objects. The geometrical descriptions of planets, stars, galaxies and galaxy clusters is utterly inconceivable without this number.

213

Without "*e*", it's hard to imagine the description of wave processes, differential equations, complex numbers, quantum mechanics and a huge number of laws we use daily. This number illustrates that an infinite series of numbers can be finite, and our brain is capable of dealing with infinities as something ordinary.

The most astonishing thing is that both these numbers are used to describe random events. In probability theory and quantum mechanics, the most popular formula looks like this:

$$\frac{1}{\sigma\sqrt{2\pi}}e^{-\frac{1}{2}\left(\frac{x-\mu}{\sigma}\right)^2}$$

where the parameter μ is the mean value, and the parameter σ is the standard deviation. Here and in a few other places later, we remove the multiplication sign "×" between the variables to make our writing compact. As we can see, this formula contains "π". In this expression, "*e*" is the exponential function, $\exp(x) = e^x$, where "*e*" is the base of natural logarithms.

This formula is called the "normal distribution". This function is also known as the "Gaussian distribution" in honor of the German mathematician Carl Gauss (1777 – 1855), who first developed the two-parameter exponential function in 1809 in connection with the study of errors in astronomical observations.

The shape of this distribution resembles a bell, with a peak in the center and symmetric sides. I illustrate this distribution in <u>Figure. 9.4</u>. It is remarkable that this function also appeals to infinities; if a certain quantity is the sum of many random contributions, independent of each other variables, and each contributing a small amount relative to the total sum, then the distribution of such a quantity with an infinitely large number of terms tends towards the normal distribution.

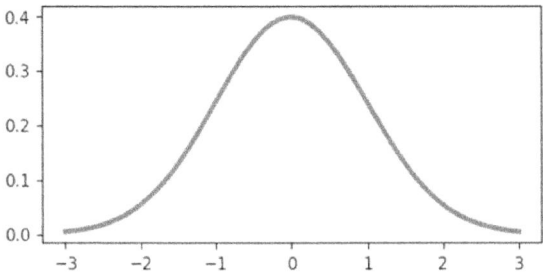

Figure. 9.4. *The drawing shows a normal distribution or Gaussian distribution, with parameters $\mu = 0$ and $\sigma = 1$.*

Isn't this surprising? Is it a coincidence? Spheres, infinities, and the description of random events "converged" in one abstract expression. And they are united by the two most important constants – "π" and "e".

This amazing observation is related to a joke (Ball and Coxeter 1987). Many years ago, Augustus De Morgan (1806 – 1871), the great Scottish mathematician and the first president of the London Mathematical Society, was explaining to an actuary (a specialist in insurance mathematics) the probability that a certain part of a group of people would remain alive at the end of a given time. He quoted an actuarial formula, including "π", which, as he explained in response to a question, means the ratio of the circumference of a circle to its diameter. His acquaintance, who had been listening to the explanation with interest until then, interrupted him and exclaimed: *"My dear friend, this must be a mistake. What does a circle have to do with the number of people living at a given time?"*

I don't know whether it was because of this remark or not, but Augustus De Morgan became very interested in spiritualism phenomena. He began to explore clairvoyance and conducted investigations of paranormal phenomena at his home. The results of these investigations were later published by his wife. De Morgan believed that his career as a scientist could suffer if he revealed his interest in spiritualism (Nelson 1969). His book was published anonymously under the title "From Matter to Spirit: The Result of Ten Years' Experience of Spiritual Manifestations".

If we pretend that there's nothing special about the Gaussian distribution, then the answer could be as follows: The description of any roundness always requires the number "π". And the sums of infinite series must somehow be described by the number "e". Both of these features define the mathematical formula of the Gaussian distribution.

But the reason for wonder is – why exactly so? There are many other formulas, especially in the theory of probability, where neither "π" nor "e" are involved. Why did the "king" of all probability formulas, the most widespread formula in the description of random processes, combine the two most common constants? What is the probability of such a coincidence?

There are about 20 forms of probability distributions. They can be bell-shaped, but not necessarily "round" as in the normal distribution. Many of them have the constant "e" within their definition. Suppose all these formulas have "e" (which is not entirely true, since there are distributions with a bell-shaped curve, like the Cauchy distribution, but they do not contain "e" inside). What is the probability that "e" and "π" would randomly "unite" with each other in the most important distribution, such as the normal distribution?

This is akin to the calculation of throwing a ball (representing the number "π") into 20 random holes and expecting the ball to end up in the hole you chose in advance (i.e., the normal distribution). Such a probability would be $1/20 = 0.05$ (or 5%). Here, we have assumed that all holes already have the number "e" inside.

A probability of 5% is not that small but also not that large. In everyday life, when we make decisions, it is reasonably small. How can we imagine its smallness? Maybe like this: If you have at hand 20 bell-shaped distributions, and you decide, at random, that one of them should represent random and independent phenomena, then the probability that it contains the number "π" would be just 5%. This distribution must contain roundness, as the number "π" appears almost everywhere there are circles, balls and spheres.

There's another interesting property of the Gaussian distribution shown in <u>Figure. 9.4</u>. We do not know how to integrate it and obtain a finite expression. It turns out that the Gaussian function does not have an elementary indefinite integral (integration of a function without any limits). Of course, we can numerically integrate it on a computer for some specific ranges or look up the answer in a table, but we cannot calculate the integral analytically. But what is astonishing, we still can integrate this function using an infinite range of values between (- ∞ and +∞)!

$$\int_{-\infty}^{+\infty} e^{-x^2}\,dx = \sqrt{\pi}$$

We just integrated this function over an infinite range and got some finite number. Funny enough, it turned out that this number is simply the square root of "π". How did this fundamental constant end up there again? Originally, it was simply obtained from the ratio of the circumference of a circle to its diameter!

Perhaps, the Universe itself adds some mysticism and says: Look, here you have a function that describes the contributions of an infinite number of random events. It's the most significant function in the theory of probability, describing how small parts of something contribute to a single outcome. Also, it's the most special function that contains the two most important constants, unrelated to probabilities,

217

without which the world could not be built. But its peculiarity is that you can calculate its area only using a computer. However, as soon as you want to use infinite intervals for integration, there's no problem at all. Our brain can handle such unimaginable operations with infinities without significant intellectual challenges. And, somehow, the obtained answer is again related to "π".

Infinite streams of random events, probabilities, spherical geometry ("π"), infinite series ("e"), and the computer — everything converges in one expression. If someone or something decided to give us a hint about the structure of the world, the most logical thing would be to link the numbers "π" and "e" in one expression.

As we showed earlier, there's a good chance that these fundamental numbers are not an infinite sequence of random numbers. They contain "markers" that may indicate encoded information. If there's someone or something who created everything, it's possible that they hid clues in these fundamental constants for those who seek.

9.5 Everything comes together

And now, I will show you something that will ignite your imagination. Look at the famous identity named after the Swiss, German and Russian mathematician Leonhard Euler (1707 – 1783):

$$e^{i\pi} = -1$$

where "e" to the power of "i" (the imaginary unit) times "π" equals - 1. We have already discussed imaginary numbers in chapter 8.5 The reality of imaginary numbers. In this formula, all fundamental quantities appear at once! Somehow, mysteriously, the constants "π", "e", the imaginary unit and the ordinary -1 create an identity. It turns out, these four concepts of mathematics are closely interconnected.

This mathematical construction has been dubbed "the most beautiful equation in mathematics" (Wells 1988). As Keith Devlin, a British mathematician and popular science writer, wrote,

"Like a Shakespearean sonnet that captures the very essence of love, or a painting that brings out the beauty of the human form that is far more than just skin deep, Euler's equation reaches down into the very depths of existence".

If we let our imagination run wild, we could say this: The identity connects infinity (the number *"e"*, see chapter 9.2 The number *e*) with sphericity (the number *"π"* in chapter 9.1 The number *π*) and the quantum world (or quantum mechanics, *"i"* — the imaginary unit) with the number 1 — representing our world. The number 1 has a unique significance: It is the first natural number after zero and can be used to construct all other integers, and thus to count anything in our world. In the philosophy of ancient Greece, 1 is considered the highest reality and the source of all existence and the basis of all numbers. How to interpret the minus sign is up to your imagination. The minus signifies something opposite.

Euler's identity can even be written like this:

$$1 = |e^{i\pi}| = |\pi^{ie}|$$

where the number *"π"* (sphericity!) is raised to the power of the imaginary unit and *"e"* (infinity!), and everything is taken as an absolute value.

For people seeking meaning in mathematical symbols and believing that the world is created as an abstract concept in the language of mathematics before manifesting it in particles and physical fields, such identities offer much room for imagination. Their decryption might sound like this: On the other side of our reality (the number "-

1"), there exists an infinite number ("*e*") of spheres ("π") of quantum worlds (the imaginary unit "*i*"). This phrase seems as if taken from books popularizing the quantum world of particles. Yet, we obtained it from the most beautiful equation in mathematics proposed in the distant 18th century.

9.6 Strangeness on unimaginable scales

In our world, there exist a few numbers that are used to describe sizes of the most important things and phenomena. Among them are the sizes of a proton, DNA, biological cell, human, Moon, Earth, Sun, Solar System, Galaxy and our Universe. Any coincidences among these numbers should arouse interest because the number of possibilities for coincidences is limited.

In chapter 8.6 Again about information, we talked about an incredibly small quantity for describing quantum effects, called the Planck constant. It is denoted by the letter "*h*". From this constant, we can derive the "Planck length", using two other fundamental constants — the speed of light and the gravitational constant "*G*". The speed of light (electromagnetic waves) in a vacuum is approximately 3×10^8 meters per second (m/s). It is denoted by the Latin letter "*c*".

The Planck length is equal to:

$$l_p = \sqrt{\frac{hG}{2\pi c^3}} \simeq 1.6 \times 10^{-35}\,\text{m}$$

For us, this is an unimaginably small distance. It has a value of the order of 10^{-35}, representing a watershed between classical physics, that is, the world of humans, and the quantum world. This is the scale at which all existing theories cease to work. Science does not know what can happen beyond the Planck length. As I mentioned before, it can be assumed that this length reflects the physical size of the smallest information quantum.

This quantity can be illustrated as follows: the diameter of human DNA is approximately 2.5×10^{-9} meters (or 2.5 nanometers). The diameter of the observable Universe is approximately 8.8×10^{26} meters. The ratio of the diameter of the Universe to the diameter of DNA is:

$$8.8 \times 10^{26} \text{ m} / 2.5 \times 10^{-9} \text{ m} = 3.5 \times 10^{35}.$$

This is how many DNAs are needed to cover the diameter of the Universe. In terms of magnitude, this number is very similar to the ratio of some usual human scale (say, average height of a man, 1.7 m, for example) to the Planck length:

$$1.7 \text{ m} / 1.6 \times 10^{-35} \text{m} = 1.1 \times 10^{35}.$$

The exact height of a human is completely insignificant to make our point, given the inconceivably large scales of these quantities.

Here we see again an amazing coincidence in both meaning and magnitudes of values. The diameter of the Universe measured in the DNA sizes (carrying biological information) is about the same as the size of a human (or some human-scale thing) measured in terms of the smallest length in physics (carrying the smallest information quantum?). Indeed, this looks very strange. There is no need to worry about the apparent difference between 3.5 and 1.1 – on the scale of this huge magnitude (10^{35}), this difference is completely negligible.

One might object to this astonishment like this: "Well, humans use a lot of various numbers. One can always find some numbers with similar coincidences!". But this is our main point: The number of possibilities, i.e. the number of the most important sizes to describe things in the Universe, is limited. We only care about coincidences which are logically related, at least when we are only dealing with the most essential for humanity numbers. See the beginning of this chapter. If you find 20 most important numbers reflecting sizes, from the size of protons to large-scale objects in the Universe, then you will be very lucky. But such a small number of fundamental sizes cannot lead to logical relationships in such coincidences, as we just illustrated, due to blind randomness. Technically speaking, the statistical sample of such numbers is too small for pure chance. Somehow, the scales of the Universe, including those of biological systems like humans, reveal some hidden structure or correlation pattern. It is not inconceivable to think that the hierarchy of sizes is arranged according to some incomprehensible logic.

9.7 The fine structure of the world

In addition to the Planck's constant "h", there are also other very small parameters for describing the microworld, such as:

- ϵ - the elementary electric charge ($1.602176634 \times 10^{-19}$ C)
- ε - the electric constant ($8.8541878128 \times 10^{-12}$ $F \times$ m^{-1}).

All of them are well measured experimentally. After their numerical value, we wrote their physical dimension (C and $F \times$m^{-1}). Almost all physical quantities have dimensions because we measure them in some units of measurement. Usually, this is some letter that follows the magnitude itself. For example, the length of something is measured in meters (m). Planck's constant "h", "e" and "ε" also have more complex units of measurement involving energy (see above). Here we will not decode physical dimensions F and C, as it is not very relevant.

On the other side of reality lies the speed of light $c \approx 3 \times 10^8$ m/s. It is used to describe cosmic objects and distances. It can be said that it separates our familiar world of humans from the world of cosmic scales.

Finding a physical constant without a unit of measurement is difficult. However, such a constant exists: It is the "fine-structure constant", usually denoted as α. It was proposed by the German theoretical physicist Arnold Sommerfeld (1868 – 1951). The constant α is written as follows:

$$\alpha = \frac{\epsilon^2}{2\varepsilon hc}$$

This value describes a quantum effect — the strength of interaction between two electrons of an atom as a result of the exchange of the virtual photons (quanta of light). The fine-structure constant also determines the size of atoms. Change this number by even one percent, and you change the Universe! For example, if you increase this number, protons will repel each other so strongly that atomic nuclei will not be able to hold themselves together. Our Universe would cease to exist.

The expression defining the fine-structure constant may seem trivial — you can construct any combination of 4 constants. But there is a "but". Actually, three:

1. α has no dimension. That is, it is simply a constant, like "π" or "e", discussed earlier.
2. α very often appears in the description of the laws of physics.
3. Its magnitude is (approximately) $\alpha \sim 1/137$

Or, to put it simply, $1/\alpha$ is approximately equal to:

137.

Isn't it fascinating that incredibly small quantities describing the microworld, and an incredibly large number ("c") for cosmic scales, "canceled out" the scales of their values, and formed a dimensionless quantity "α" (that is, without units of measurement)? Moreover, they created an easily recognizable whole number with which people deal in everyday life.

Debates about what this feature reveals about the Universe have been ongoing for nearly a century. Wolfgang Pauli joked that his first question after death would be — "Why 1/137?". He wrote:

"I believe that in the natural sciences, a different, opposite approach will emerge, which will be connected with ancient mystical foundations".

American physicist, Richard Feynman, said:

"It's one of the greatest damn mysteries of physics: a magic number that comes to us with no understanding by man. You might say the 'hand of God' wrote that number, and 'we don't know how He pushed his pencil'. We know what kind of a dance to do experimentally to measure this number very accurately, but we don't know what kind of dance to do on the computer to make this number come out, without putting it in secretly!"

It should be noted that this number can be measured in experiments. It turns out that $1/\alpha$ is somewhat larger than what we initially wrote. Specifically, it equals:

$$137.0359990.$$

Does this fact "destroy" our expectation that this number should be integer? I think, no. Firstly, theoretical physics and mathematics have repeatedly proven that beauty in the mathematical expressions almost always leads to the correct answers, regardless of the small deviations from the experimental measurements from the expected values. Secondly, it is possible (or even almost certain) that this number can change slightly over time and different energies. In 2010, the VLT telescope received indications that this constant might change over time. However, there are no confident confirmations of changes in the fine-structure constant (Berengut et al. 2011). From the perspective of modern quantum mechanics, the fine-structure constant changes depending on the energy scale of the interaction. Thirdly, if there is a small deviation from the beautiful mathematical construction, it may indicate a violation of some symmetry. That is, the original system had a value of 137 at some energy scale of interaction, but over time (or under the influence of certain factors), it was violated, and a deviation appeared. And, finally, I would like to see more measurements of this constant. The experimental value of this constant is almost always dominated by one experiment. In the history of physics, there have been situations where the appearance of a new method of measurement or a new experiment fundamentally changed the numerical values of physical variables.

You might ask: What's so unusual about a constant being so close to a whole number? There are many important constants, and some of them are bound to be close to integer values. As I've mentioned before, we know relatively few dimensionless constants in physics. The fine-structure constant is the king of all constants. According to some estimates, life is only possible (Barrow 2001) when $1/\alpha$ is within the range of 85 to 180. If you were to randomly select a real number from this interval, the probability that such a random number would also be close to some integer number is 0.069, assuming that the degree of closeness will be as high as for the $1/\alpha$ constant. See our calculations in chapter 20.11 Appendix. This isn't a very small probability. It's roughly equivalent to getting heads four times in a row when flipping a coin. But, for decision-making, it's a sufficiently low chance. If you knew that the value of a company's stock would go down with a 6.9% chance, would you invest your money in buying such stocks? I think yes. And that's the reason for the surprise. Why do physical constants on completely different scales create a value close to a whole number?

It's interesting that observing a small deviation from 137 significantly reduced the number of scientists searching for meaning in the fine-structure constant. But this is not well justified. What if α is more fundamental and precisely equals 1/137, as it reflects some symmetry that was violated or a change in conditions over time? In this case, perhaps one of the constants making up the expression for the value of α should differ from the magnitude expected from experiments before this symmetry was violated. For example, the electric constant could have differed in the past from the value that is known today.

Nevertheless, the pursuit of beauty in describing the world has led to some unexpected findings. It turns out that $1/\alpha$ can simply be derived from the number "π":

$$\frac{1}{\alpha} = 4\pi^3 + \pi^2 + \pi$$

Here, a surprise awaits you. This expression quite accurately reflects the experimental value of this constant. Compare these three values shown below:

- Using ϵ, h, ε, c constants $\quad = 137.036010952$
- Using expression with c and π $\; = 137.036303776$
- Experimental measurement $\quad = 137.035999206$

The first value is obtained using the physical constants "ϵ", "h", "c" and "ε". The second value is derived using an expression from the numbers "π". And the third one is obtained from experimental data. You can check this yourself using the program in chapter <u>20.11 Appendix</u>.

How is it possible that a simple manipulation with the number "π" leads to a value that is so close to the measured value of the parameter $1/\alpha$? This astonishing connection between the number "π" and the fine-structure constant was noted a long time ago (Roskies and Peres 1971). As we've mentioned, we can confidently say that every time we see the number "π" in a mathematical expression, it always indicates that the expression describes some form of sphericity. There are even attempts to explain the value of the fine-structure constant through the formula based on the value of "π" only, using a simple geometrical model with spheres (Yee 2019). It turns out, $1/\alpha$ can be obtained thanks to the relationships with spherical geometrical shapes as a result of the movement of something filling empty space. Note that these shapes do not have a definite size. They can be infinitely small or infinitely large. As we remember, the fine-structure constant describes a quantum effect — the strength (or probability) of interaction between two electrons because of the exchange of virtual photons between them. But what does this have to do with geometry?

Indeed, one could try to go through many different expressions with the number "π" and find the value of $1/\alpha$. However, you won't be able to find such a simple and elegant formula as the one shown above. If we believe in a certain beauty of mathematics, such a connection with geometry might not be coincidental. From the history of science, we know that simple and beautiful expressions almost always correspond to some description of the reality of this world.

Here, I allow myself some figurative digression. If there really is a connection between the energy emitted by an electron during a photon emission and the motion of something within certain geometric spheres, then we might be observing some kind of "mechanism" that underlies the Universe. This is exactly how the physicist-inventor Nikola Tesla (1856 – 1943) envisioned the Universe. In our example, it's some kind of "clockwork mechanism" where incredibly small, spherical ("π") gears mesh together and strike sparks of light ("α") in the form of photons!

And now, a little joke from somebody who came up with this whole "mechanism". Take the first 6 digits of the number $\pi=3.141592$, square them, and then add them up. Let's try:

$$3^2+1^2+4^2+1^2+5^2+9^2+2^2 = 137.$$

Coincidence? Why exactly 6 digits of the number "π", and not 7 or 8? The reason is that this "trick" is for us — humans. It uses the decimal system, created with a "tie" to the biological features of humans. In the past, people created this system using ten fingers. But why the number 6? We also discussed this in chapter 3.10 Remarkable number. The number 6 appears in chapters 9.1 The number π and 6.6 The Kaiser and the War and many other places. Remember, according to the Bible, the number 6 is related to the creation of man. And according to science, the number 6 is "woven" into all steps of world and life creation. Who knows, maybe the whole Universe started with this joke? And then the value of the fine-structure constant was assigned to 137? Why does the specific value of this number for the microworld matter, if it looks so wonderful and mysterious?! I think such beauty will also be appreciated by anyone who seeks meaning in everything.

9.8 The infinite beauty of fractals

What would you say if you see an incredibly complex drawing with various protrusions, peaks and depressions, that upon magnification roughly repeats the pattern, regardless of how many times you enlarge it? To be more precise, we can enlarge this picture an infinite number of times, spending our entire life on this activity, but we will never reach a limit. The pattern of peaks and valleys will appear similar at any magnification, although not exactly repeating the previous "landscapes". Such visual structures are called "fractals".

This is what you see in <u>Figure. 9.7</u>. This drawing was made using a computer. In mathematics, such a complex mathematical construction is called the "Mandelbrot set" fractal (Mandelbrot 1983).

But first, a small digression. When I started writing this chapter, I tried to insert a caption under this computer-generated figure. I typed the word "Mandelbrot". As soon as I did, my wife, who was in another room at the time, said the name "Benoit Mandelbrot". It turns out she was reading aloud the epigraph of Nassim Taleb's book "The Black Swan" (Taleb 2008). The full phrase in that book was like this: "Dedicated to Benoit Mandelbrot, a Greek among Romans".

I was stunned. What? I had just written that word! I can say for sure that I had never uttered this surname in my wife's presence, although I had dealt with fractals many years ago when I was still a student. We decided to check — was this the same professor Benoit Mandelbrot who was one of the first to use computers for visualizing fractals? We checked this name in an encyclopedia — yes, it was him. My wife had never heard of fractals or Mandelbrot before. When she said this phrase, she completely randomly opened a book that she had borrowed from a library because someone had recommended this author to her. The book had no direct relation to fractals.

I must say right away without any calculations: The likelihood that two people pronounce such a complex surname as "Mandelbrot", and at the same time have never discussed this topic until this very moment, is absolutely tiny.

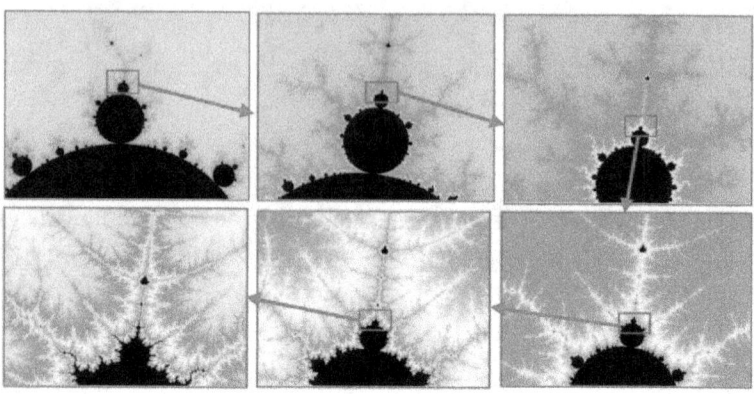

Figure. 9.7. *Here is shown a fractal called the Mandelbrot set. Each of the images was obtained by magnifying a small part (inside the red rectangle) of the previous image. It is usually said that such structures are "self-similar" upon magnification.*

But let's return to <u>Figure. 9.7</u>. Such complex geometric structures are usually characterized by "fractal dimensions". They provide an estimate of the fractal's complexity. Even simple geometric shapes are characterized by such dimensions. For example, a point has a fractal dimension of 0. A line has a fractal dimension of 1. Further, the number 2 corresponds to surfaces (having length and width), while the number 3 is intended for sets describing volume. Unlike simple geometric forms (point, line, etc.), the fractal coefficient can have a non-integer value. This means that fractal sets fill space differently or sparsely compared to simple geometric shapes.

Coastlines, like many other geological features, have a similar fractal structure when viewed from different distances as their scale changes. They too can be characterized by a fractal dimension. For example, the coastline of Australia has a fractal dimension of about

1.14. (Husain, Reddy, and Sajid 2021). The fractal dimension of the coast of Great Britain is 1.25.

To describe such incredibly complex patterns that look just as complex after magnification as they do before, an enormous amount of information is needed. Creating the Mandelbrot fractal at all scales requires an infinitely large amount of data. However, this rule does not apply to the abstract world of mathematics. The Mandelbrot set can be created using a simple recursive relation that generates the necessary data through a computer simulation:

$$z_{n+1} = z_n^2 + b, \qquad z_0 = 0$$

This set consists of a complex number "b" for which z_{n+1} remains bounded. All you need to do is iterate this expression using a computer. To create a Mandelbrot set image, you must take the numerical value of each point (or pixel) and run it repeatedly through this formula. Then, the obtained result should be plugged back into this mathematical expression. And what comes out of it, insert again, and so on.

But let's not delve into the mathematical details. You've just seen something incredible: An infinitely complex structure can be generated using a surprisingly simple and finite mathematical expression. Only quite recently, with the advent of powerful computers, have mathematicians realized how rich the world of such simple formulas is.

If this world really was created, then the problem of the huge variety of details that we see around us could be precisely solved thanks to such compact mathematical structures. In religious teachings, infinity is not a problem for God, even if he desires to see all the pixels created within the fractal. But even for an all-knowing being, it's not necessary to construct an infinite number of small details and observe every created point from the boundless expanse. It's sufficient to create a simple mathematical structure that will generate an infinite number of details of our world like an autonomous "factory" of things.

Can it be said that the numerical data generated by iterating a recursive mathematical formula, which creates an infinity of beautiful patterns, constitutes some kind of information? Generally speaking — no. These are just data that lead to a certain regularity, which we perceive as beautiful patterns after we create a graphical representation of these data on a computer. See chapter 3.2 Information is easily recognizable.

However, if someone already knows everything about these data, how they look and what patterns they can lead at any step of magnification, then the situation changes fundamentally. These chaotic data become information. And the economical way to obtain them is to use iterations of the recursive relation on a computer. As we can see — the presence of consciousness once again becomes central to the definition of information.

There is another conclusion that can be drawn. For an omniscient being that possesses all the information about our Universe, a simple set of rules can be used to iteratively create all the smallest details of our world. Or it could create an infinite number of universes, making small changes in a certain mathematical algorithm. Here, one can cite examples of elementary cellular automata, i.e., programs with very simple rules capable of creating infinitely complex data with some order (Wolfram 2002).

But our world is not just fractal geometry. It is filled with events. Often — random, often — not. What relation do fractals have to the events that occur around us?

It turns out that many distributions of particles in nature can also be described by fractal parameters. Imagine two high-energy particles accelerated to speeds comparable to the speed of light in an accelerator, like the one used at CERN. Then, they collide together in a particle detector. Hundreds of other particles arise as a result of the collision. Or imagine another scenario where an energetic particle from space collides with oxygen molecules in our atmosphere, creating many new particles. All particles produced because of such collisions create statistical distributions. Their shape strongly resembles fractals or the coastlines of Britain or Australia. These particle distributions are also described by non-integer fractal dimensions. Such dimensions have been well measured for different colliding particles at different energies (Kittel and De Wolf 2005). I was dealing with such questions during my student years.

Here, random probability distributions and geometry (more precisely — fractal dimensions) have met again. Something similar was seen in chapter 9.4 Infinity, sphere and randomness where the number "π", describing sphericity, "penetrated" the description of probabilities of an infinitely large number of events.

9.9 The Great Pyramid and numbers

We previously discussed the primary rule for identifying noteworthy coincidences: We must focus on significant objects and events. When discussing buildings, the most important numbers should reflect their geometric dimensions and geographic locations. But the buildings themselves must also be highly significant. If the size of one house on a street matches the size of another house on a neighboring street, there's nothing unusual about it. There are too many houses among which there will be many similarities. It's just statistical noise,

233

thanks to chance. And the statistical sample or the number of possibilities is enormous.

However, if we inquire whether there's a coincidence in the most important characteristics of a building that is the largest and oldest in the world, then the situation changes. Circumstances become worthy of investigation when we find a match in the dimensions of the building with something that is equally significant to people. For example, with the dimensions of the Earth, Moon, or Sun. Or if the location of the building somehow coincides with fundamental constants, of which there are not so many.

This is precisely the case with the Pyramid of Khufu, located on the Giza Plateau in the vicinity of the Egyptian capital. This structure is the tallest of all pyramids and the only one of the "Seven Wonders of the World" that has survived to this day. It cannot be denied — this is the most important structure from which the search for unexpected coincidences begins. Historical records detailing the exact construction methods of the Great Pyramids of Giza have not survived, adding to the mystery surrounding their creation.

I've mentioned before that once a person realizes they're dealing with some sort of wonder or anomaly, additional circumstances almost always emerge to amplify that feeling. And this sense of the unusual can serve as a starting point for calculating the probabilities of such events occurring by blind chance. Indeed, many intriguing coincidences are associated with the Pyramid of Khufu. For example, its geometry includes the number "π" (see chapter 9.1 The number π). The ratio of the Great Pyramid's base to its height is approximately 3.1425. The precision of the "π" value at the time of the pyramid's construction is remarkably good. But let's consider just two coincidences unrelated to the number "π".

234

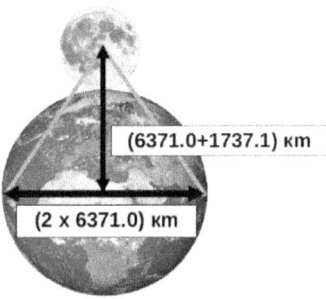

Figure 9.9. *Perhaps this is how the ancient builders imagined the proportions of the Great Pyramid of Giza.*

Here is one of them: The original height of the pyramid was 146.6 meters, and the base size was 230.0 meters. The ratio of height to base is 146.6 / 230.0 = 0.637. Now, imagine you place the Moon on the Earth and mentally draw a triangle with the Earth's diameter as its base. See <u>Figure 9.9</u>. The average radius of the Earth is 6371.0 kilometers (km). And the height of such a triangle is the distance from the center of the Earth to the center of the Moon. The average radius of the Moon is 1737.1 km. Let's calculate the height of this triangle relative to its base:

$$(6371.0 \text{ km} + 1737.1 \text{ km}) / (2 \times 6371.0 \text{ km}) = 0.636.$$

We've obtained the agreement with the ratio (0.637) of the dimensions of the Pyramid of Khufu within 0.17%. Notice that it doesn't matter in which units we measured it, as we're interested in the ratios of numbers. The approximate sizes of the Moon and Earth have been understood since antiquity, but there is no historical evidence supporting such stunning precision during the construction of the Pyramids of Giza.

What's the probability of such a coincidence? Assuming that the height of the pyramid could vary within the range of 100 to 150 meters while keeping the base fixed, the probability of a random co-incidence with a value of 0.636 within a relative error of 0.17% (or less) is 0.005. This calculation is detailed in chapter 20.12 Appendix. A value of 0.5% is sufficiently small for a random coincidence. It's equivalent to the chance of flipping a coin 7 to 8 times in a row, with each time landing heads (or tails). Of course, this probability depends on the chosen height interval of 100 to 150 meters. Our assumption is that this pyramid should remain impressive, not less than 100 meters in height. Building a pyramid taller than 150 meters would be more challenging. Since the calculation program is provided in chapter 20.12 Appendix, the reader can recalculate this probability by using a different height interval.

A skeptic might say that one can always find some numbers and choose the ones that match. That's not the case at all. We only used the most fundamental quantities, such as the dimensions of the building and the most important characteristics of our main celestial bodies. There aren't many of them, around 10 or even less. Secondly, there's the rational association in this coincidence. Specifically, there was an attempt to create a hypothetical 2D projection of the pyramid onto a triangle mentally, using the sizes of celestial bodies, placing the Moon on the Earth and drawing three lines through their centers.

Let's consider another coincidence related to the location of this pyramid. The modern value of the speed of light in a vacuum is c=299,792,458 m/s. The coordinates of the Great Pyramid: 29.9792458°N (north latitude). Amazing, isn't it? Even if a similar number is taken from the range 29.9802000°N — 29.9782000°N that can be attributed to the pyramid (Benedictus 2020), this does not make this coincidence less surprising.

Because probabilities multiply for independent events (see chapter 3.3 About probabilities), we're talking about a very small probability that these two independent coincidences (in the size of the pyramid and its position coordinate) are possible by blind chance.

Let's dive deeper into the aspect of the coincidence in the position. The unit of measurement for the speed of light is defined in meters. However, it's noteworthy that the meter itself wasn't defined until the year 1771. The speed of light, with the precision I've mentioned, wasn't known until 1975. How could this coincidence have appeared in ancient times?

Let's suppose the Great Pyramid was built by an advanced civilization. How would they reason? Firstly, this coincidence exists precisely in the decimal system, based on the biological feature of humans (10 fingers on both hands). This would have been a brilliant insight for the pyramid builders. They knew the speed of light with very good precision and translated it into the decimal system. The value of latitude was likely borrowed from Babylon, where the number 60 was used for calculations. See chapter 3.10 Remarkable number. This system was applied to a circle with 360 degrees. And 60 divisions were used for minutes and seconds. Therefore, they could have solved the latitude problem. But how did they know that the meter would become the primary unit of measurement for humanity in the future, and that we would use a method of dividing circles borrowed from Babylonian astronomy? Does this mean they knew the future?!

If you're trying to find an explanation for the extremely low probability that these coincidences are possible by chance, and the hypothesis of aliens who were aware of our civilization's future is too fantastical, then there's one way out. To solve this problem, our hypothesis, which runs through all the sections of this book, will come in handy. I want to make it clear right away that it's no less fantastic than aliens who knew the future of our civilization and created all these puzzles. Here it is: Perhaps the Great Pyramid is some kind of beacon left to us from the distant past. Or it's an instrument of cali-brating retrocausality, which we've discussed several times in chapter 6 Examples of significant coincidences and 8.2 Nonlocality. Like a "sponge", it absorbs our knowledge, which, as it accumulates, is re-flected in the past of this structure. This is how it goes: Initially, a pyramid was created. It astonished people with its size. But there were no coincidences in the numbers representing its size or position. The pyramid influenced the emotional states of people who expected addi-tional coincidences to confirm its uniqueness. This changed the past to meet those expectations. When people measured the number "π" with sufficient accuracy, they began to look for it in the geometry of this structure. And they found it, because the past "satisfied" their ex-pectations. The pyramid began to appear with dimensions in which the number "π" was present with the necessary precision. When people measured the sizes of the Moon and Earth, they started looking for these dependencies in the pyramid, hoping to confirm its uniqueness. The past changes again, meeting such expectations: Now the Great Pyramid includes this change in its size. When people measured the speed of light with exceptionally good accuracy, the search for addi-tional uniqueness was incorporated into the information about this pyramid. Now the coincidence involving the speed of light and lati-tude appeared. And so on.

It should be added that according to our hypothesis, it is we ourselves, through our consciousness, who make these changes in the pyramid. If our consciousness is connected to the world of abstract forms, for which there is neither time nor space, then such a hypothesis is quite justified. For abstract forms, the Great Pyramid does not exist in the past, just as it doesn't for us.

According to this hypothesis, the entire immaterial world of the past undergoes changes due to the connection of our consciousness with the timeless world of ideas. All we have is the fleeting moment of the present. All the past is elastic. It changes with rare bursts of consciousness at key moments of our present. But the effects of editing the past are much harder to detect for insignificant ancient objects due to greater random noise and less of our attention. This is particularly true for structures that lack substantial size and historical solidity akin to the pyramids at Giza.

I don't rule out at all that other mysteries of deep antiquity can be explained by retrocausality. For example, polygonal masonry of polygonal stones or the Nazca Lines in the territory of modern Peru could be the result of such a phenomenon. They are artifacts of a transformed past, where everything is constantly changing. We don't need aliens from other planets or advanced civilizations of the past (Hancock 1995) to explain them. All of this was done by us, through our connection with the reality of meaning.

10 Premonitions of the future

One day, while driving home from work, I noticed a flock of Canadian geese flying. They spanned across most of the sky. There wasn't a single cloud, so every tiny detail of this flock could be clearly seen. The geese were flying in a V-formation because such flight reduces wind resistance and helps the geese save energy.

It was January 10, 2024. Winter was already in full swing, but there was no snow. The temperature was relatively high for Chicago this time of year — about 50°F (10°C). Just two days ago, my wife and I had been feeding a large group of geese at the pond near our house, where they felt completely safe and comfortable. However, now, as I was approaching my home, they had all disappeared. I decided that they had most likely joined the flock I had seen 10 minutes ago.

But why had they left? The geese should not have migrated during such a mild winter. The most interesting fact about the Canadian geese living in the suburbs of Chicago is that they make migration decisions to the south (to places like southern Illinois, Tennessee, and Arkansas) based on many factors. Often, they do not fly away at all if the winter is mild and there is little snow.

I only understood this three days later. Chicago was hit by a snowstorm, followed by a very sharp drop in temperature to -4°F (–20°C). How did the geese "know" about this change in the weather three days before the storm? Did they perceive information that was not available to us? They certainly didn't read the weather forecast.

Assuming that the geese set off on their journey at some random time during one winter month, coinciding exactly with the beginning of a sharp change in weather by 30 degrees, then such a probability would be less than 1/30. Of course, it was not a pure coincidence. Science has already established that animals can sense the weather and make decisions based on such premonitions. They do not need great intellect for such decisions. Instincts work exceptionally well (see chapter 3.7 Brain and information).

Nevertheless, migrations of birds are still a phenomenon that we do not know much about. Where do birds "learn" that it's time to embark on their long flight? We can't even fully understand how animals navigate in space, or where birds get the ability to find their way back home.

We also know little about why birds (and other animals) decide to migrate and how they find the right routes. While humans can create particle colliders to understand the structure of matter at scales of 10^{-18} meters and build spacecraft that fly beyond the Solar System, we still understand relatively little about quite ordinary things. For example, the behavior of animals that we see every day in their natural environment remains a mystery. How do birds sitting in front of our windows decide to fly thousands of kilometers with precise navigation during flight and return?

Animal migration is usually considered instinctive behavior, passed down from generation to generation. These are inherited behavior patterns. What individual animals learn is passed onto future generations. It is believed that instincts are encoded in DNA. Instinctive behavior patterns are not learned, as they are the result of evolutionary processes and are inherited. How instincts are encoded in DNA is not yet known, but it is an active area of research.

In the case of Canadian geese, perhaps an instinct for certain weather phenomena, such as changes in pressure and temperature, etc., was triggered. For example, a study by ornithologists at the University of Western Ontario showed that birds can detect changes in air pressure, allowing them to anticipate storms (Boyer 2019). This is a mechanically pre-programmed response to external factors. Such sensitivity must be highly developed, as their lives directly depend on these survival skills. Most people have lost such abilities. Our lives do not depend much on the weather, and our intellect requires extensive knowledge to make decisions.

Undoubtedly, instincts are ingrained in us from birth for daily, or often repeated situations. For example, the instinct for seasonal migration, maternal instinct, hunting instinct, etc. However, in the case of rarely repeated or special situations, on which life or death depends, instinct is an unlikely explanation. In the case of the Canadian geese, where the decision to migrate south was made three days before a sharp change in air temperature, the situation remains puzzling.

The only other option might be that information about an upcoming rare cataclysm, upon which life directly depended, was received in real-time from an external source. Instincts perceived it and prompted the geese to act.

Humans do not have innate needs for behavior programs. We use intellect most of our time. Here, I want to talk about a case that does not fit the definition of an instinctive reaction to a future event. In the 2010s, I often traveled from Chicago to CERN (Switzerland) to work on the hadron collider, which accelerates high-energy particles to near the speed of light using electromagnetic fields. On one of the trips, I mechanically swiped my Belarusian passport. Until that trip, I had only taken my U.S. passport, which allowed me to enter the Schengen Area countries but not Belarus, where my parents lived. Upon arriving in Geneva, I learned that my father had suddenly passed away. The next day, I flew from Geneva to Minsk with my Belarusian passport. Had I only had my U.S. passport, I would not have been able to enter Belarus and say goodbye to my father. Before this event, I had traveled to CERN about ten times, using only my U.S. passport. Thus, the probability that I accidentally put the necessary passport in my bag, considering the number of trips over the previous years, is 1/10 (10%). While this is not a very small probability, I was still struck by such a coincidence.

Of course, this event is not related to an instinctive reaction. It simply could not have developed during the evolutionary process. In this case, absolutely nothing indicated the possibility of such a coincidence.

In this chapter, we will examine several cases that are more convincing than the one described above.

10.1 Nostradamus

Michel de Nostredame (1503 – 1566), a French apothecary, better known as Nostradamus, is considered one of the most famous seers of the 16th century. Interpreting his predictions is quite a complex matter. It is also not easy to assess the probability of their occurrence due to random coincidences.

However, in some cases, it is possible to make some approximate estimates of the role of chance in his predictions. This applies to cases where Nostradamus mentions specific numbers and places. For example, in his book "Les Prophéties" from 1555, Nostradamus wrote (Century I, Quatrain 51):

"London shall demand the blood of the innocent, burning in the fire in 3×20 + 6 (=66). Many palaces shall be destroyed, and the old lady shall fall from her great height off her throne".

As you may have noticed, the number 6 is significant in this prediction (see chapter 3.10 Remarkable number). An intriguing fact is that in 1666, the Great Fire of London truly devastated the city. The disaster destroyed almost 90 percent of the homes. More precisely, nearly 13,000 homes and 90 parish churches were burned to the ground. The fire also threatened Whitehall Palace, where the English queen's residence was located.

To estimate the probability that Nostradamus could predict such an event, if we consider Europe only, it would be reasonable to assume that he should think about a fire in any of the 10 largest cities in Europe. Of course, the number 10 is an assumption, as there were

245

many more large cities. But let's assume there are only 10. Hence, the probability of such a prediction is 1/10 = 0.1.

Further, we can estimate the number of possible disasters which can be relevant for London. Given its location, we can consider three possibilities: fire, flood or epidemic. It's hard to imagine other tragedies, such as an earthquake, if we talk about London. Therefore, this probability will be set to 1/3.

And most importantly, he mentioned the number (time) 66. Although he did it in a peculiar way, writing the time as 3×20+6 = 66. It seems he meant a century within a millennium. Clearly, this isn't a day of the week or a month. Thus, the probability is 1/100 = 0.01. However, we do not know for which century this prediction was made. In principle, he could be talking about five possibilities — 1566, 1666, 1766, 1866, 1966. Let's stop at 1966, as 2066 has not yet come. Therefore, we will increase the obtained probability by 5.

Thus, the probability that Nostradamus randomly guessed the Great Fire of London is:

$$0.1 \times 1/3 \times 5 \times 0.01 \approx 0.001666.$$

This is a sufficiently small value. It roughly corresponds to the probability of flipping "heads" (or "tails") 6 times in a row when tossing a coin. One cannot help but notice the presence of 6, and the year of the tragedy, 1666, in the numerical value of the probability. We deliberately truncated the infinite number of sixes in our record to make this numerical coincidence look amusing. However, we did not intentionally adjust our line of reasoning so that 1666 would appear in the probability value. The number 6 simply "dances around" this event! But I just want to caution you in the fact that we used 10 random cities; it could be 9 or 11, which would slightly change this probability value.

For me, this prophecy is the most interesting because it precisely indicates the number and location of the event. Many believe that Nostradamus predicted many other important events. In this book, we will focus on the most well-documented prophecies that include specific numbers and names.

One of the predictions is directly related to Nostradamus's profession — he was a physician. Therefore, a meaningful connection based on the type of activity should exist. Recall that we used this principle in chapter 6 Examples of significant coincidences, where we showed that coincidences among people of the same profession (politicians, physicists, etc.) occur much more often and are more vivid. It is the meaningful or logical connection that serves as the unifying principle among people in such groups. Here is one of such predictions by Nostradamus:

(Century I, Quatrain 25): *"The lost thing is discovered, hidden for many centuries; Pasteur will be celebrated almost as a God-like figure; This is when the moon completes her great cycle; But by other rumors he shall be dishonored."*

Surprisingly, there's a coincidence here too. Louis Pasteur (1822 – 1895) was a French chemist and microbiologist who discovered the principles of vaccination, microbial fermentation, and pasteurization. He also proved that bacteria do not appear spontaneously, as was previously thought. Instead, they grow from already living organisms in a process called "biogenesis". Although Pasteur was not the first to propose the "germ theory", he convinced much of Europe of its validity. He invented the process of removing bacteria — named "pasteurization" after him. His early work also led to the creation of vaccines for rabies and anthrax. According to science historian Gerald L. Geison (Geison 2014), Pasteur used his competitor's discoveries to make his anthrax vaccine functional. This discovery partly "disgraced" the great scientist, as Nostradamus predicted.

We won't delve into whether Pasteur was disgraced or not. He indeed discovered what had been "hidden for many centuries". This serves as a quite interesting analogy for microbes. Of course, there could be several other interpretations, not necessarily related to microbes. The most interesting part of this Nostradamus prediction is the surname. It is a relatively rare surname, and guessing such a surname is not easy. Certainly, it does not fall within the 3600 most popular French surnames ("3600+ French Last Names"), so the probability of guessing it is less than

$$1/3600 = 0.00027.$$

At the same time, a parallel interpretation is also possible. It could talk about a pastor or shepherd (if we translate "pasteur" from French), who rediscovered lost knowledge hidden for centuries, but then his discovery was questioned. In this case, there was no prediction. At least not on a scale comparable to the discoveries of Louis Pasteur. But the very combination, where a pastor or shepherd makes a discovery, appears quite absurd.

As with any explanation, raw data always requires interpretation using some belief system. Nostradamus's predictions appear precisely as data that needs to be processed using some established logic system. In our reasoning, we assume that there must be a logical connection, meaning Louis Pasteur's discoveries were directly related to Nostradamus's professional activities as a physician and chemist. I don't think a similar discovery by a pastor or shepherd would have evoked an emotional connection in Nostradamus. We used such a system of reasoning for scientists and politicians in chapter 6 Examples of significant coincidences.

And now, let's assess the probability that the two predictions described in these quatrains were nothing but coincidence. As we remember, we need to multiply the probabilities. If we assume that Nostradamus somehow randomly guessed both the Great Fire of London and the name of Louis Pasteur at the same time, then such a probability:

$$0.001666 \times 0.00027 = 4.45 \times 10^{-7}.$$

Of course, this is a very small probability for a correct guess.

It's often said that Nostradamus couldn't have predicted anything because there are too many interpretations of his 942 poetic quatrains, supposedly predicting future events. Suppose that each of his predictions can be interpreted in five different ways (though I've heard about several possible interpretations). Then the probability that the prophecies in his manuscripts are a matter of random coincidence and pure guessing:

$$(942 \times 5) \times 4.45 \times 10^{-7} = 0.0021.$$

This is a small probability for random chance (0.21%). It is approximately equal to the likelihood of getting "heads" when flipping a coin 9 times in a row.

You might ask, why are all these predictions so complicated, convoluted and easily interpretable in various ways? Here, one can adopt the viewpoint of Carl Jung. He believed that information about the future could "leak" out only in the form of certain hints and signs. Or in the form of scattered data fragments that are "distorted" by the brain. After all, its role is to exist in a world of dense matter, changing in space and time. It is not "tuned" to perceive all the nuances of the informational environment where neither space nor time exists. Jung said (Jung 1973):

"Unlike the psychic phenomena that are open to perception, .. the collective unconscious... cannot be directly perceived or 'imagined', and due to the 'unimaginability' of its nature."

We will return to this issue in chapter <u>16.4 Symbols and forms of ideas</u>, where we will explain the viewpoint that some access to that inaccessible reality might be possible through symbols.

10.2 The Urantia Book

Once, while visiting a library in the suburbs of Chicago, my gaze fell upon a thick book titled "The Urantia Book". It seemed to have garnered little interest, as it was placed on the very top shelf. It was the most imposing book in size on the entire shelf, comprising about 2,000 pages. Quickly flipping through it, I realized it used many scientific and religious terms. The author was not listed. This book piqued my interest. I took it home to read and to delve into the internet to find out who wrote it.

It turned out, "The Urantia Book" (Urantia Foundation 2008) has a religious character. It contains documents that are presented as the fifth "epochal revelation", according to the statement of anonymous authors. It provides an in-depth description of the Universe's structure. It delves into topics of humanity, the divine, and the life of Jesus Christ. The Urantia Book describes the destiny of mankind and teaches that faith is the key to spiritual progress and eternal survival. It also outlines God's plan of progressive evolution for human society and the Universe as a whole. Urantia is our planet, Earth, according to this text.

The authors of "The Urantia Book" are unknown. It was first published in Chicago in 1955. However, the main text (comprising about 200 papers) dates back to 1934 - 1935. It is believed that the authorship belongs to one of the patients of the American psychiatrist William Sadler. According to Sadler's accounts, this patient occasionally fell into a type of sleep during which he spoke on behalf of various supernatural beings. However, after awakening, the patient remembered nothing about it. The name of this patient was not disclosed. How "The Urantia Book" was written (or completed) between 1935 and 1955 is unclear. Subsequently, this book served as the foundation for the esoteric movement "The Urantia Brotherhood". It is considered that the movement was influenced by the theosophy of the Russian religious philosopher Helena Blavatsky (1831 – 1891).

How was "The Urantia Book" written? Judging by the text, there must have been a rather large group of enthusiasts with education in various fields of science. They would have had to use typewriters, as computers did not exist at that time. But most importantly, they would have had to keep their activity secret for their entire lives.

I was particularly interested in the vast amount of material on topics of interest to science. Since it was claimed that these revelations were received from "celestial beings", I hoped to find scientific facts that humanity learned about after the book's publication. Indeed, it contains a tremendous amount of information on the structure of the world, astronomy, physics, and chemistry.

Naturally, I decided to figure out how the knowledge about the world, conveyed to us by "celestial beings", differs from what we know in the 2020s. It turned out that these beings didn't know much. At least, what they knew, people already knew or speculated about in the 1930s to 1950s.

For example, I attempted to find in the pages of this book how elementary particles, such as protons that make up atomic nuclei, are structured. It is these protons that are accelerated at the Large Hadron

Collider at CERN. The word "proton" in Greek means "first", and the name for the hydrogen nucleus was given by Ernest Rutherford (1871 – 1937) in 1920. The Urantia Book mentions protons multiple times. However, protons had already been discovered when this book was being written. Unfortunately, the book said nothing about the structure of protons. This isn't surprising if the text was created by a human. However, if this book was written under the dictation of some cosmic beings, they should have precisely known what a proton consists of. As we know today, this particle is made up of three quarks. They are structureless (point-like) particles. There are 6 types of quarks: up, down, charm, strange, top, and bottom. Each quark flavor also has three "colors" associated with it (red, green, and blue), according to the theory of quantum chromodynamics. Besides these quarks, there are also numerous gluons and quark-antiquark pairs. Quarks and gluons cannot leave the proton due to the phenomenon of confinement, or the impossibility of obtaining free-state quarks. Scientists only learned about this in the 1960s and 1970s.

Looking back, we now understand that the scientific information in the book generally aligns with the science of the 1930s and 1940s. Nonetheless, it contains information that could be regarded as predictions. It turned out that there are scientists who have attempted to compare the scientific descriptions of the Universe from "The Urantia Book" with modern knowledge. The book contains 31 predictions on various scientific topics. They can be grouped into the following categories (Ginsburgh and Taylor 1987) (Taylor 2017):

- 15 predictions that align with modern science,
- 10 predictions that partially align with science,
- 6 predictions that do not match science.

The fact that 48% of the predictions align with science is quite an interesting observation. Nevertheless, I would have expected more insight from "celestial beings".

And yet, I've been pondering the following: If there truly exists a way to obtain new information, such a channel must be full of information "noise" that distracts from the useful messages. A source of uncertainty is necessary, allowing for doubt about what is written in the book and leaving ample room for critique. This loophole is for skeptics. It's essential for preserving our free will (see chapter 17 Free will) and motivation to seek scientific truths.

Let me elaborate on this. Imagine a scenario where "The Urantia Book" had predicted with 100% accuracy all the natural laws discovered by the 2020s. Naturally, this would halt natural scientific progress. All that would be needed is to find patients with similar symptoms and use them for additional information about the structure of the world. Why educate students or invest in scientific facilities? But most importantly, it would immediately limit the range of our freedom and possibilities for creating anything new.

If all the scientific facts were fully confirmed, then all the spiritual and religious information in this book would have to be absolutely correct as well. Indeed, the real value of this book lies precisely in its spiritual and philosophical guidance. It affirms the existence of God, describes the purpose of humanity, and then provides the most detailed biography of Jesus Christ, elucidating the meaning of his teachings. This implies that Christianity is the true religion. All we need to do is to behave as this teaching expects from us. This fact would nullify all other religions that have found their own paths to understanding God utilizing their unique historic and cultural heritage. We will return to the issue of freedom of will later.

Well, after examining "The Urantia Book", I thought — this book does not contain many predictions — only 31 in total. If there really is a glimmer of new knowledge in this book, there should be one or two predictions that are 100% accurate. Even if this entire book wasn't written under the dictation of cosmic beings, the people who wrote it must have been under some psychological effect. After all, not everyone can decide to create 2,000 pages of complex text while not believing in what is written. Could it be that some true and precise knowledge found its way into the minds of the book's authors? And not just precise — its appearance should be highly improbable for the 1930s - 1940s.

It turns out that the book indeed contains a few scientific predictions. For instance, it talks about "dark islands of space", which will soon be discovered by humans:

15:6.11 (173.1) *"Burned-out Suns. Some of the dark islands of space are burned-out isolated suns, all available space-energy having been emitted. The organized units of matter approximate full condensation, virtual complete consolidation; and it requires ages upon ages for such enormous masses of highly condensed matter to be recharged in the circuits of space and thus to be prepared for new cycles of universe function following a collision or some equally revivifying cosmic happening"*.

This description matches what we understand as "black holes" today. These are regions of space where a huge amount of mass is packed into a tiny volume. This creates a gravitational pull so strong that not even light can escape it. Mostly, they are formed by the collapse of giant stars. Even before "The Urantia Book" was written, Albert Einstein had completed his work on the theory of gravity, within which the German astronomer Karl Schwarzschild (1873 – 1916) in

1916 calculated the properties of space and time inside and outside a collapsing star.

In 1967, theoretical physicist John Wheeler (1911 – 2008) described the properties of black holes, using this term in science for the first time. Later, his work was continued by Stephen Hawking. Interestingly, it was Wheeler who, in 1990, hypothesized that information is a fundamental concept in physics. In 2015, physicists discovered gravitational waves and concluded that they were generated by two black holes in the final seconds of their merger to form a more massive black hole. This observation definitively confirmed the existence of these objects. However, it's worth noting that candidates for black holes were observed as early as the early 1970s (Cygnus X-1, a galactic source of X-ray radiation). As we see, the text of "The Urantia Book" aligns quite accurately with the concept of black holes.

My attention was drawn to another prediction. In physics, there is a particle named neutrino, which participates in weak interactions with matter. The neutrino was first proposed by Austrian-Swiss physicist Wolfgang Pauli (1900 – 1958), whom we've already mentioned. In 1933, he pointed out that detecting such a particle would be a very difficult task. Thus, the neutrino was not experimentally discovered until 1956. For a long time, it was believed that neutrinos had no mass. Here is what "The Urantia Book" says in several places in the text:

41:8.3 *"In large suns — small circular nebulae—when hydrogen is exhausted and gravity contraction ensues, if such a body is not sufficiently opaque to retain the internal pressure of support for the outer gas regions, then a sudden collapse occurs. The gravity-electric changes give origin to vast quantities of tiny particles devoid of electric potential, and such particles readily escape from the solar interior ..."*

42:8.5 *"When atoms perform radioactively, they emit far more energy than would be expected. This excess of radiation is derived from the breaking up of the mesotron "energy carrier," which thereby becomes a mere electron. The mesotronic disintegration is also accompanied by the emission of certain small uncharged particles"*.

The text containing sentences about "tiny uncharged particles" closely resembles descriptions of neutrinos. It's possible that the author of the book already had this information as a theoretical idea from Wolfgang Pauli. Furthermore, according to the book, these particles have a very small mass (Taylor 2017). It was only in 1998 that scientists confirmed neutrinos indeed have a tiny mass.

There are also several other interesting scientific descriptions that align with modern views. However, it's difficult to determine whether they are based on hypotheses existing in the 1930s and 1940s or not.

But let's shift our focus from science and look at the historical events contained in this book. The most astonishing thing about this text is that it includes one of the most detailed descriptions of the life of Jesus Christ. So detailed, in fact, that it describes many events with specific months and days of the week. These are not predictions based on logical reasoning about a particular scientific topic and are not prophecies about the future. Therefore, such information is quite harmless in terms of its impact on the future. For me, it was a "hook" to check how much sense there is in the description of these historic events. If the exact date of an event is known, then it is possible to calculate the day of the week and compare it with the text. I selected 14 events for which the time was precisely indicated (including day, month, and year) and calculated the day of the week. Then I compared them with the text. Out of 14 events, only one day of the week did not match. The calculations are presented in chapter 20.13 Appendix.

Note that the text uses the Julian calendar. Unfortunately, many events did not specify the year, and I did not include them in the calculations. It is possible that these years could easily be deduced by reading the book more carefully, but I did not have the time for that.

What is the probability of guessing 13 days of the week correctly? The answer: $(1/7)^{13} \approx 10^{-11}$. This is an incredibly small probability, even if this book was wrong for one or two days. A similar accuracy check was also conducted by followers of the Urantia society, which claims that 30 times of events described in this book were verified. All of them turned out to be true.

This tiny probability confidently affirms that the book was written by people with a good understanding of mathematics. The calculations for the 1930s must have been exceptionally monotonous, especially considering that the book contains about 100 dates of some events. For those who believe that the book was dictated by higher beings, this is yet another reason to believe in the description of the life of Jesus Christ.

Another possibility is to assume that this book is a fraud. Someone wrote a book on spirituality, being not at all spiritual. Obviously, the author must be a knowledgeable scientist (or a group of scientists). But why? If they believed in the written text, then they must have understood that they were committing a sin by making up this text. Why risk their soul if they fervently believe in its existence? They were clearly very erudite and spent, presumably, a lot of time on large monotonous calculations without a computer. They were clearly not insane. They wrote 2,000 pages of very complex text for... fame and money? This is unlikely, as the book itself is distributed for free, and the authors' names are not listed. Undoubtedly, in the times of aggressive materialism in science at the beginning of the 20th century, scientists with religious worldviews were not welcomed. Perhaps this explains the complete secrecy in writing the book.

It's up to you to decide what this book means. I think the truth always lies somewhere in the middle. Perhaps, a patient of psychiatrist Sadler could have said a lot of this information under hypnosis, which seemed mysterious and intriguing to Sadler. This information was documented. Then, these records inspired other scientifically-minded people, who supplemented the themes of Sadler's patient's thought fragments.

I believe this book contains unusual information that deserves attention. Maybe this information does not come from cosmic beings but from the deep layers of intuition that draw islets of knowledge from the common information field. Who knows?

The book contains several quite interesting predictions about the structure of the microworld and the Universe, which I have not touched upon. Let's wait and see if such predictions come true. I think the Urantia Book is indeed useful and can perhaps have an impact on the spiritual growth of society.

10.3 Predictions and checks

Working on my book, I became convinced of how difficult it is to find predictions that can be unequivocally characterized as having come true. More often than not, I could not find original documents that confirmed prophecies. Mostly, these were post-predictions, that is, when news of the prophecy appeared after the events themselves had already occurred. In many cases, the clairvoyant readings were so vague that it was very hard to determine what they actually meant.

There are many prophecies in history, but most of them can be considered as foresights based on knowledge and experience. For example, Nikola Tesla (1856 – 1943), an engineer and inventor, predicted in 1909 that someday people would walk around with phones in their pockets. In a New York Times article, he wrote that

"It will soon be possible to transmit wireless messages all over the world so simply that any individual can carry and operate their own apparatus".

The American clairvoyant, Andrew Davis (1826 – 1910), predicted the widespread appearance of internal combustion engine cars in every detail in 1856. But prototypes of such engineering designs already existed, and it could be guessed that it was precisely such cars that would replace steam cars. In these cases, these were creative foresights. People used analytical thinking, experience, and good knowledge of their professional field.

Finding true predictions was so difficult that I even thought of a fantastical scenario: What if correct predictions existed at the time where they were made, but this information was removed by synchronicity later to avoid any paradoxes, to allow for a natural flow of history? Supposedly, synchronicities work outside time and space, so reconfiguring the past and veiling texts with predictions should not have been a problem. Maybe all these clairvoyants did indeed make predictions at the time, and therefore they rightfully earned their fame. However, all the true predictions that touched on serious events were "diluted" with some "informational noise", that is, unfulfilled prophecies, or even destroyed, so that humanity would not pay attention to real foresights and would continue to develop according to its natural laws.

Certainly, the reasons could have been more mundane. There was no internet in the past. Predictions were mostly made in local newspapers or on the radio, if books were not written. This is why finding original material is very difficult. And, of course, there could have been plain fraud.

I reviewed the predictions of the religious philosopher Helena Blavatsky (1831 – 1891), the American clairvoyant Andrew Davis (1826 – 1910), the mystic Edgar Cayce (1877 – 1945), the Soviet artist Wolf Messing (1899 – 1974) and the Bulgarian clairvoyant Vanga (1911 – 1996). According to the internet, all of them made many false prophecies. Their foresights of new technologies can be explained by the fact that prototypes of such technologies already existed at the time. But what's interesting is that the primary sources with their records were quite hard to find. Mostly, I found interpretations of their predictions by third parties.

10.4 Premonitions of tragedies

In this book, we are interested in predictions that are difficult to reconcile with subject knowledge. We are interested in the intuitive forecasting of the onset of upcoming tragic events. Such types of foresight mostly arise in the consciousness as "glimmers" and "flashes". Because these events are usually associated with strong emotional experiences, they are capable of affecting our connection with the primal source of information beyond the constraints of time. Here, possibly, we deal with spontaneous access to new information about the future. Rational thinking is not applicable in such situations.

As I've mentioned before, when discussing predictions made by individuals, the first thing that stands out is how difficult it is to find well-documented predictions before the event has taken place. The case with Nostradamus is special, see chapter 10.1 Nostradamus. He was one of the few who wrote a book with many potential predictions.

One of the most impressive cases of well-documented prediction I discovered on a shelf in a Barnes & Noble bookstore in 2024. The book had a striking bright orange color and was titled "End of Days: Predictions and Prophecies about the End of the World" (Browne and Harrison 2008). The main author of the book was the American writer, medium, and psychic Sylvia Browne (1936 – 2013). The book was published in 2008. I was intrigued by one prediction that read like this:

"In around 2020, a severe pneumonia-like illness will spread throughout the globe, attacking the lungs and the bronchial tubes and resisting all known treatments," ... "Almost more baffling than the illness itself will be the fact that it will suddenly vanish as quickly as it arrived, attack again ten years later, and then disappear completely."

Of course, she was referring to the Coronavirus (COVID-19), which gained strength in 2020. What is the likelihood that such a prediction was a mere coincidence? We will make a rather conservative estimate: Assume that the writer was looking at the time span from 2008 to 2019. This gives a 1/(2019-2008) chance that the timing was guessed correctly. There are several ways to talk about a disease. Let's assume that a deadly disease could only affect 5 major organs: the brain, heart, stomach (and associated organs), central nervous system, and lungs. This gives a 1/5 chance for guessing the type of disease. Since she emphasized that the disease would quickly fade away, we'll assume a probability of 1/2 (50% for quickly and 50% for slowly). The total probability:

$$1/(2019 - 2008) \times 1/5 \times 1/2 = 0.0092.$$

This is the probability that Browne randomly guessed when and what the global disease would look like. For all practical purposes, the probability of 1% is quite small. It's roughly equivalent to getting 6-7 heads (or tails) in a row when flipping a coin. The psychic made several other predictions for the 2020s that turned out to be incorrect. But even if she made 10 unsuccessful attempts, thus increasing the obtained probability by an order of magnitude, we still obtain a good accuracy – 10% If a new outbreak of disease with similar symptoms appears in 2030, as Browne predicted in 2008, the probability of a random guess would be exceptionally small (just multiply 0.0092 by 1/(2030-2008)).

This is one of the rare cases where the predictions were indeed well-documented, and the book itself is easily purchasable even now. However, in the vast majority of cases, finding specific predictions in publications is not straightforward. Here are a few examples of predictions that are difficult to verify, as the predictions themselves were not documented. Unfortunately, I could not find their primary sources.

In May 1969, Joseph DeLouise (1927 – 2006), a Chicago psychic and author known for his predictions of future events, predicted a crash of a jet plane near Indianapolis (a city in the USA). The prediction was presumably first made on the radio in Eddie Hubbard's WLS program and reiterated when he was a guest of radio host Bob Allard on WOC in Davenport, Iowa. He mentioned that the number 330 would be significant but did not know in what way, and that 79 people would be killed. Indeed, on September 10, 1969, at 15:30 (3:30pm), an Allegheny Airlines DC-9 aircraft collided with a private plane near Indianapolis. The crew of four and 79 passengers died.

This prediction appears quite astonishing and accurate. Even a rough estimate of the probability that such a prophecy was a random guess gives a value of 10^{-5}. However, all available sources describing this prediction were created in newspapers and on the Internet after the tragedy occurred. It was impossible for me to find recordings of that radio program. Therefore, I cannot assert that such a prediction was made before the tragedy.

Here's another prediction. The Bulgarian clairvoyant Vanga (1911 – 1996) stated in 1980 that "*At the end of the century, in 1999 or 2000, Kursk will be underwater, and the whole world will mourn it*". Indeed, at the end of the summer of 2000, the Russian nuclear submarine "Kursk" sank in the Barents Sea. The tragedy claimed the lives of 118 crew members. However, this prophecy does not have documentary confirmations.

I decided to dig through Internet forums to see if there was any material proof that Vanga made such a forecast. It turned out that there were many people who assured me that they had read this forecast in Soviet newspapers 20 years before the accident. There could be no mistake. Those who lived in the city of Kursk, located in the central part of Russia, remembered this article well because Vanga paid attention to their city. At that time, it was quite honorable when she was mentioning someone's city in her predictions. She was indeed very famous. But in the 1980s, this prediction seemed absurd: If it was about the city of Kursk, then the "flooding" of this city was hard to imagine. This city is situated on high ground. One of the versions was that such a prophecy appeared in some local newspaper as a joke. But absolutely no one on these forums could find that old newspaper. It should be noted here that there were few newspapers in the USSR. They were carefully scrutinized for disinformation, religious content and pseudoscience of any sort. However, the official press often wrote about Vanga. She enjoyed the discreet backing of the top party leadership.

The most "mundane" explanation in the Kursk story is this: Finding an old Soviet newspaper where Vanga made the prediction is indeed quite difficult for various reasons. But one can assume that clairvoyance was indeed present. In this case, we are dealing with a rather accurate prediction. The probability of a mere guess is exceptionally low, considering the nature of the tragedy (sinking) and the word "Kursk".

Another explanation is the "Mandela Effect". This phenomenon is associated with false collective memory. It is assumed that the Mandela Effect happens when many people have memories that contradict actual facts. In this case, there was no prediction. In the 1980s, Vanga said nothing about Kursk. All these people on the forums are mistaken. They have false memories.

Here's another example of a possible Mandela Effect: People vividly remember how the USSR General Secretary, Nikita Khrushchev (1894 – 1971), banged his shoe on the podium and threatened the USA with the words "we will show you Kuzma's mother" at the UN meeting in 1960. This is an expression of an unspecified threat, similar to "teach someone a lesson". You can even find a photo of the general secretary with a shoe in his hand and his mouth open. But nothing of the sort happened. He simply banged his shoe but did not utter such a phrase. The phrase "We will show you Kuzma's mother yet!" was actually spoken by Khrushchev during a visit to the American exhibition in Sokolniki (Moscow) in 1959.

There are various explanations for the Mandela Effect, ranging from the most fantastical — like a glitch in the computer simulation of our world, to the splitting of universes with different timelines. One of our fantastical hypotheses is the editing of the past. More precisely, the events occurred as people remembered them. But the information that has reached us has changed. In the case of Vanga's prediction, it can be assumed that all these people from the city of Kursk remember the article about the flooding of Kursk correctly. Such an article indeed existed in the 1980s. But all material evidence has been destroyed, thanks to synchronicity. Why? To minimize the impact of such predictions on the natural course of events.

If you read modern articles (including Wikipedia) about Vanga in the 2020s, it turns out that she did not make any predictions. She was just a fraud working for the secret service of the Communist Party of Bulgaria. But how to explain her influence in those years? There is clearly some contradiction here. Could it be that right now, there is a falsification of truthful information about that time, so that no one doubts that predicting the future is impossible?

As we can see, individual forecasts that lack credible archival records before the tragedy occurred are not very reliable. However, it's possible to analyze statistical data about people's reactions before the tragedy happened. In other words, instead of looking for predictions before the events happened, one can focus on analyzing data after they occurred. If people can sense impending doom, they should unconsciously avoid situations where their lives are threatened.

For example, one could check how many people were in cars involved in accidents and compare that to the average number of people in cars that were not involved in accidents. If people unconsciously sense that traveling is unsafe, they might skip the trip. In the case of cars, such an approach would depend on too many extraneous factors. For instance, drivers with more people might be more distracted by conversations, which means there could be more accidents in cars with many people.

Or another method is to check how many passengers were on airplanes during aviation disasters compared to similar flights before the disasters occurred. This is also not easy, as aviation disasters are quite rare, and there are long intervals of time between them.

However, there is one dataset that looks very interesting. In the 1960s, Chicago was one of the main railroad hubs. More railroad tracks extended in different directions from Chicago than from any other city in North America. Many people used trains to travel from one place to another. Train crashes also happened quite frequently, as the railway system was overloaded and there were no computers to alert and manage movement in the dense web of railroads. This time was ideal for conducting an analysis of the number of passengers in trains before and during rail disasters.

This was indeed done in 1956 by researcher E.W. Cox, who wrote an article for the Journal of the American Society for Psychical Research. He sought evidence of foresight into deadly train crashes involving 10 or more fatalities by studying the number of ticket bookings in trains during and before the tragedy. He only considered specific trains and made calculations for 28 accidents.

His hypothesis was this: Any form of foresight could be detected by finding a decrease in the number of ticket bookings for specific trains on the days when accidents occurred. The results of such research were indeed positive. He discovered that on days that ended in disaster, there were slightly fewer passengers traveling. He interpreted this fact as an unconscious motivation of people to make decisions that would allow them to avoid negative consequences. However, the effect was not strong enough to prove such a phenomenon definitively. Chapter <u>20.14 Appendix</u> presents these data and plots a graph of deviations from the average.

It should be noted that, recently, a considerable number of similar studies have emerged. These studies indeed lead to the conclusion that such a phenomenon exists (Ben et al. 2016). In most cases, critical comments on such conclusions are driven by a different belief system, rather than a scientific analysis of these results. These remarks are not based on attempts to replicate similar experiments and publish refutations. This indicates a tendency to ignore data, if it doesn't fit to pre-defined expectations. As such, for science, this issue remains open. But for those who believe, based on their personal experience, this question has long been resolved.

11 Dreams and coincidences

Every day our consciousness "shuts off" for a sufficiently long period of time. We do not hear, do not see, and do not feel. Our consciousness is as if in another world. Waking up in the morning, you might ask the question – where have I been all this time? And if you suddenly wake someone up at night, they won't immediately understand which world they are in.

Currently, science already knows quite a lot about what happens in our brain during sleep. But as is often the case in science, it can only describe what happens to the body during these times. However, the reasons why we sleep are still not fully understood.

I have always been interested in this question. If we are in a position of natural selection occurring in nature, where the most adapted animals survive, then sleep is the most inconvenient thing that can happen to a biological being pursued by hunters. There is no protection from predators during sleep. In this case, animals with the least time needed for sleep would gain an incredibly large advantage over animals with more hours of sleep. And after hundreds of millions of years of evolution, one might expect that only organisms that have learned to manage without deep sleep, or have learned to rest in a half-sleep, would become the dominant species on Earth. If this has not happened, then there are several possibilities: The mechanism of natural selection has problems, or sleep is such an important mechanism for life that complex life without it is unimaginable. One of the possible explanations suggests that sleep helps the process of removing harmful by-products of life that have accumulated in the brain cells.

Let's assume that sleep is an absolute necessity for the functioning of complex organisms. It's needed for the body and brain to rest and cleanse the system. But then, what is the purpose of dreams? For experts, this question remains unresolved. The prevailing theory is that dreams consolidate and analyze memories. After such analysis, dreams serve as "rehearsals" for various situations and problems that a person encounters during the daytime. When we sleep, our brain continues to operate. During sleep, restorative processes occur in the body. These are operations of consolidation and discarding unnecessary information.

11.1 Dreaming as information

The average person experiences about four dreams per night. On average, people dream about 2 hours every night. In most cases, people dream during the phase of rapid eye movement ("REM") sleep, during which cognitive processes occur. However, this cognitive process is entirely different. We do not think in thoughts and words, as we do when awake. When we dream, we have no concept of time and no linear cause-and-effect relationships. Instead, we think and feel this world through symbols, emotions and signs. Possibly, it is the language of signs that has more direct access to the world of idea-forms (see chapter 16.4 Symbols and forms of ideas).

And here, a simple question arises: Why do we dream during the processing and analysis of memories? It is a highly energy-consuming process, as energy is spent on dreaming. After all, constructing meaningful scenic images and new situations is mentally laborious. In fact, we are not resting while dreaming, as the brain activity is almost the same as during wakefulness. The brain simply operates differently (Borbley 1988).

Even if you are not familiar with the biological processes occurring in the brain, you can notice one peculiarity. Processing information and simultaneously watching a dream are quite incompatible operations. It's as if you are making a backup of data on a computer, moving and processing a huge amount of information, and at the same time, you turn on a movie on the same computer for viewing. You are not just watching this movie – you are creating it from the same fragments of information that are being deleted or compressed in size. Is it not obvious that such operations with data are usually done at night when the computer is not very active?

Another problem is that over their lifetime, a person sees about 130,000 dreams, and almost all of them are not remembered.

One of the rare individuals who could remember dreams was Nobel Prize laureate in Physics Wolfgang Pauli, whom we mentioned earlier. Carl Jung collected about 1,300 of Pauli's dreams and analyzed them. Subsequently, Jung used this information in his works "Psychology and Religion" and "Psychology and Alchemy". The main idea of his approach is that in dreams, you encounter archetypes, that is, universal symbols representing fundamental human experience and the collective unconscious. Our dreams are a world of symbolic archetypes into which we inadvertently peek, disconnecting our mind and senses from the external world.

But if we forget the vast majority of our dreams, then what is their benefit? Here lies a double paradox. First, the brain expends energy creating dreams at a time when these resources are most needed for processing information. And then, all this information disappears once the individual forgets the dream. Moreover, it is remarkable that the dreams that are remembered, in most cases, are not ones that are useful during wakefulness. For example, I have never noticed a situation where I could use my dreams for any purpose. However, recalling dreams during the day, I experienced emotional excitement: The fragments of dreams that were remembered were too unusual and could not be described in the concepts we apply in this world.

Here, an analogy is again appropriate. Suppose every night you go to the cinema and watch several films in one go. Your body is relaxed and resting. You are in the darkness. Then, you leave the cinema and find that almost everything you saw has been completely forgotten. And these films themselves are never repeated again because they are all destroyed. You may even find that some of your friends have not left this theater: When you are immersed in the world of a film, you are so absorbed by it that you become easy prey for predators. This is what should happen in nature with sleeping animals.

You've probably noticed the absurdity of explanations for the existence of dreams. The expenditure of energy for creating symbolic data, perceived by our consciousness as a dream, and then the destruction of this information does not fit into a logical explanation. It might be assumed that dreams are useful because, in some way, they change our emotional attitude to life at the subconscious level during wakefulness. But this is not very convincing. Nor is it convincing that dreams are just some byproducts of information processing, such as compression and deletion. I think you will never find examples where creation of meaningful information is an unintended result of data processing for a totally different purpose.

A much more logical situation would be one where dreams simply should not exist, so that the organism could use its biological resources to process memories at 100%. And if dreams are an unintended byproduct of some data transformation, then they should consist of fragments of information from previous days. This last scenario is also quite rare.

However, there is a more logical hypothesis, albeit somewhat fantastic. As we said earlier, a logical and elegant explanation is most likely a reflection of reality. This explanation is as follows: We expend energy on dreaming because it is part of our life program and a piece of the puzzle as to why we are here at all. Sleep is the moment when we connect to the unconscious, to a subjective psychic reality that lies beyond the physical world. Why? Perhaps, to exchange information with this unconscious. Then, dreams are erased because they are not needed during the day. Our physical world is completely different from the world of psychic reality. This information is utterly useless in the Universe where the constant flow of time dictates its rules for physical activity of complex organisms in their "shells" made of chemical elements. And the strict cause-and-effect logic of our reality is so different that even the reflections of the world of ideas in our dreams are not something useful for us.

As we explained earlier in chapter 7.2 The Lawgiver, entirely abstract theoretical models of our world are capable of describing observed phenomena. This applies not only to mathematical equations but also to all general concepts of science. We will examine how this can occur in chapter 16.4 Symbols and forms of ideas.

In the case of dreams, such a model might look like this. Dreams are needed for the exchange and transmission of information. This is one of the purposes of our being here. It is a very important process for living organisms, even though it poses the danger of being eaten by predators during the process of sleep. This is why natural selection "stalled" at this point and did not prefer individuals who do not sleep. Or sleep in such a way that their eyes alternately open to react to the slightest movements around. These are just a few examples of scenarios where the body and brain can rest, without exposing themselves to the danger of being eaten by predators.

Perhaps being in this world is not natural for us. People in their biological suits must interact with the physical environment, like divers exploring the ocean floor. This is daily life. But, from time to time, it's necessary to surface, refill the oxygen tanks, and hand over the samples of fauna collected from the ocean floor. This is sleep. And then we put the suits back on and embark on a new journey at the bottom of the ocean. And so it goes, until the suit wears out and the moment comes when we must return to the surface permanently, giving the worn-out equipment back to the world of atoms and molecules. Or replace the suit with a new one, if you decide to return.

In this analogy, sleep with dreams is an information exchange process. Information is not specifically created for dreams. Then what does it mean to see dreams? Perhaps our brain receives fragments of information involuntarily, interpreting them as dreams? Is this a case of unintentional access to a resource, where we send our information? Here's an analogy: Suppose a movie camera records information through the lens onto film. This is analogous to sleep. But even in this process, you can look inside the camera's lens and see distorted images against the sensor of this camera. This sensor has open access to the world of meanings, where your information is directed. The distortion is introduced by the brain's intellect, which exists for very specific purposes — to interact with this world. What you see inside the lens is not newly created information. It was always there. During sleep, you enrich it with your experience obtained from this world.

As we said before, one way to test a logically beautiful and consistent model is to look at its predictions.

One of the simplest predictions of such a fantastical hypothesis is that sleep should be much deeper if you had more intense activity during the day. In this case, you collect more impressions, emotions and general information. Naturally, transmitting all these experiences will take more time. If sleep is just packaging, unpacking and data analysis, then such dreams should also take more time. In this sense, it's impossible to distinguish a materialistic description of sleep from an explanation where our sleep is just the state of the organism during which information is transmitted.

However, if sleep is a means of transmitting information during which new information can penetrate through a feedback channel and be deformed by the intellectual component of the brain, then our model will lead to a somewhat different prediction. Namely, I am referring to dreams that can describe some future events. This is because time, as such, does not exist in the reality to which we send our information.

275

11.2 Dreams and new information

So, if sleep is just a side effect occurring during compression, compaction, deletion and analysis of information, it is highly unlikely that such processes can lead to dreams. Especially if they contain events that might happen in the future.

However, if dreams are some new information deformed by our intellect, "leaked" from the spiritual world at the moment when the "gates" are open for information exchange with the reality of meanings and the universal mind, it is quite natural to expect the possibility of prophetic dreams. As we supposed that other reality does not have time, or its sequence of events directed in a completely different way, compared to our arrow of time. Thus, the "response" of future events in our dreams is possible.

This is exactly how Carl Jung perceived dreams. He wrote (Jung 1901): *"Dreams prepare, announce, or warn about certain situations, often long before they actually happen"*.

If sleep is some random process creating a "mess" from some past events or images with unconscious fantasies, then among 8 billion people, there will be those whose dreams have indeed predicted future events. This is simply the law of large numbers, multiplied by the 130,000 dreams we see in our lifetime (though not all dreams are remembered).

In this book, we will examine a story, following our principle of selecting situations based on criteria not related to gaining fame through these astonishing events. We will consider a case where a dream was clearly prophetic. This person falls into the category of individuals who became famous not because they had a prophetic dream, but because they became known for other actions.

11.3 Mark Twain's dream

For me, Mark Twain is among the ten most famous writers. In high school, I read the works of Mark Twain, which were part of the mandatory school curriculum.

This story is about a dream of Samuel Clemens (also known as Mark Twain). He dreamt about his brother Henry, whom he had arranged to work on the same steamboat "Pennsylvania" where he himself worked. One night, Samuel stayed at his sister's house. He dreamed that he was looking at a metallic coffin standing on two chairs. In the coffin lay lifeless Henry, with a bouquet of white flowers on his chest, among which was a single red flower. He was dressed in one of Samuel's suits.

Soon after that, in the real world, the "Pennsylvania" set sail for New Orleans. But Samuel got into a fight with another crew member and as a result, was transferred at his own request to another ship, leaving the steamboat "Pennsylvania". Near Memphis, Tennessee, the steamboat's boiler exploded. About 250 passengers died. His brother was also seriously injured, having been burned in the boiler explosion. A few weeks later, the brother died from an overdose of opium, which was administered to alleviate his pain.

His face was untouched, and kind volunteer women were so moved by his beauty and innocence that they gifted him the best metallic coffin. Samuel Clemens described the meeting with his deceased brother as follows:

"When I entered the funeral parlor, there lay Henry in an open metallic coffin in the middle on two chairs. He was dressed in a suit from my wardrobe. He had taken it without my knowledge during our last

stay together in St. Louis. I immediately realized that my dream, seen a few weeks before, was exactly reproduced here, at least in all these details, — and I thought one detail was missing; but it soon appeared: an elderly lady entered with a large bouquet, mostly of white roses, with a red rose in the center, and placed the bouquet on his chest".

Mark Twain continued to ponder the circumstances of the death until the end of his life. He was one of the first to join the Society for Psychical Research after it was founded in London in 1882, hoping its researchers could help him understand the mechanism of dream foresight.

Let's calculate the likelihood that this dream was a mere coincidence. Since we know very little about all possible situations in this dream, we will simply assume that the probability of each event is 50%. That is, it could happen with a probability of 50%, or not happen with a probability of 50%. We have already discussed earlier why it makes sense to do this in chapter <u>3.3 About probabilities</u>. Now, let's calculate:

- The probability that Henry is dead (or not) is 0.5
- The probability that he is in Samuel's suit (or not) is 0.5
- The probability that he is in a metal coffin (or not) is 0.5
- The probability that the metal coffin is standing on two chairs (or not) is 0.5
- What is the probability of a white bouquet with a rose in the center? Here, several color combinations can be considered: white, red, yellow — the most common colors. You can construct 6 binary color combinations and 3 combinations with a monochrome bouquet. As a result, we get the probability 1/9 or 0.11. We will assume that the probability that some flowers will be brought is 100%.

The total probability is

$$0.5^4 \times 0.11 = 0.0069.$$

This is a small probability. It is lower than the probability of getting "heads" 7 times in a row when flipping a coin.

It should also be noted that this is a conditional probability, meaning that we calculated it on the condition that Mark Twain saw only Henry, but no one else, although he had seven brothers and sisters. Perhaps, this is the most conservative probability that can be obtained, since we took 0.5 for the probability of each event. For example, the probability that his brother will be in Samuel's suit or the appearance of a rare metal coffin in a dream could be much less than 50%.

Let's now analyze this example from the perspective of our concept, discussed in chapter 5 How to crack the code. From my point of view, Mark Twain is among the ten most popular authors in the world. This dream describes one of the most significant events for Mark Twain. He was close to his brother not just because of their kinship but also because they worked together in the same place. This creates a meaningful connection and may increase the likelihood of synchronicity. Both facts significantly reduce the number of possible events when calculating the probability of coincidences due to chance.

Note that the examples of synchronicity with dreams provided by Carl Jung (Jung 1973) are not always convincing. Here is an excerpt from his book:

"I discovered 'coincidences' so significantly related, and the probability of their 'chance' was such an astronomical figure, that they were clearly 'meaningful.' As an example, I will cite a case from my practice. I was treating a young woman, and at a critical moment, she had a dream in which she was given a golden scarab. As she was telling me this dream, I sat with my back to a closed window. Suddenly, I

heard a sound behind me, reminiscent of a quiet knock. I turned around and saw some flying insect that was hitting the outside of the window glass. I opened the window and caught the creature in flight as soon as it flew into the room. It turned out to be the closest analogue of a scarab that one could find in our latitudes. It was a scarabaeid beetle, the rose chafer (Cetonia aurata), which, contrary to its habits, clearly wanted to get into the dark room at that very moment. I must admit that nothing like this had ever happened to me before or since, and the patient's dream remained unique in my practice".

I do not at all think that the probability of such a coincidence is "astronomically" small. I think that almost all of us have encountered situations where, after thinking about something or seeing something in a dream, we encounter it in real life. In this situation, the probability is not extremely small, because we do not know how many patients came to Carl Jung for treatment. We only learned about this young woman because this coincidence occurred to her, i.e. after it happened. The most important thing is that thousands of inconspicuous events happen to billions of us over the course of a day. And we usually do not pay attention to them. We start paying attention only to events that somehow coincide. But the number of possibilities is huge.

The case described by Jung is not a significant event for the fate of this patient. But if it was such for this patient and for Jung (who used this example in his book), then retrocausality could have played a role. Initially, the patient saw this beetle in the office. It was so striking that his memory of the dream was altered in such a way that the beetle appeared in the patient's dream. Naturally, this significantly intensified the moment of surprise.

11.4 From personal experience

I am sure that many readers of this book have seen dreams that can be interpreted as predictions of the future. And if you haven't seen them yourself, you have surely heard about such dreams from your relatives or acquaintances. However, calculating the probability of a prophetic dream occurring by chance is not very easy due to the need for an individual approach. As has been said, we must focus only on dreams in which the predictions concern significant, life-altering events. There are not many such events in a person's life, perhaps 10-20. For such situations, the law of large numbers ("anything can happen when there are too many possibilities") is not a big problem to ensure that such predictions are indeed unusual.

Here, I will share several cases related to dreams that can be characterized as prophetic. One of the dreams is related to the death of my mother. She was put into a medically induced coma after her clinical death and spent a long time in intensive care in Minsk. I lived near Chicago at that time. Shortly before her death, when she was still in a coma, I went to sleep at 10:00 PM and woke up at 11:00 PM. I don't remember ever waking up exactly an hour later at this time of night before. The dream was as follows: I was at an online meeting, waiting for my turn to speak. But my mother tries to distract me. I started to get angry, trying to explain to her that I am very busy right now. But she doesn't listen and keeps trying to distract me from the meeting. Eventually, I raised my voice, urging her to leave me alone because I needed to speak at the meeting. I turned my back to her and felt her hugging me from behind. I felt tenderness, love, and shame that I was not paying attention to her. I woke up at 11:00 PM in a cold sweat. According to my sister, who visited the hospital during this time, my mother never regained consciousness after that time. She died a week later. That day when I saw her in my dream was her point of no return.

Further details of such events are described in chapter <u>6.8 From my experience</u>.

Another event was related to the death of my grandmother. In the 1990s, I was studying and living in the Netherlands. For several days, I woke up precisely at 3:15 AM from a strange dream where I saw my grandmother standing in front of me, just as I remembered her from before. I couldn't figure out why this was happening to me. At that time, she lived in a village in Belarus, but I knew nothing about her condition. One day I was informed that she had died. When I compared the days when she was dying with the days when I saw her in my dream, the time matched quite well.

I think almost everyone has heard similar stories. These are typical warning dreams. Are there logical explanations for such dreams? Could it be that the brain, being active during sleep, constructs forecasts for future events? Possibly. But this hypothesis is unlikely to clarify the amazing timing coincidences of such dreams with actual events. And it certainly does not explain the enormous number of details seen in dreams that turned out to be remarkably accurate, as in the story described in chapter <u>11.3 Mark Twain's dream</u>.

12 Mycelium of meanings (story)

Assuming there's an external source of information, I don't think it permeates and "flows into" this world through some physical channels (such as undetected particles). Many of the amazing coincidences and situations described in this book arise at different time segments and are noticeable on both vast and incredibly small scales of matter, as in the case of the fine-tuning of the Universe. They affect people's destinies too. The best analogy would be a computer simulation, where we are the players, and the surrounding world is a multi-dimensional simulated world. Figuratively speaking, the material world is a drawing pattern on the canvas of meanings conveyed by designed information.

This story serves as a brief respite for those less inclined towards abstract philosophical concepts. In 2017, I had the opportunity to take a break from my daily routine in a wonderful place — the Italian Alps. On one of the rainy autumn evenings, I was walking along a winding mountain path towards my hotel. I heard the sound of steps approaching from behind. Turning around, I saw a young man who enthusiastically greeted me and offered his company. I was not averse to brightening the solitude of my journey and gladly accepted.

He was tall and handsome, but there was something surreal in the look of his dark eyes — an unusual confidence and strength emanated from his gaze. I felt I had seen him before but could not remember where. The cold wind blew his wavy hair. The setting sun, behind the foggy mountain behind us, created a halo of glow over his head, adding a touch of mystique. He introduced himself as Martin, and we continued on our way together.

Judging by the waterproof cloak and rubber boots, I had met a mushroom picker. But there was one unusual detail that seemed utterly out of place — a new shiny leather briefcase filled with freshly cut mushrooms.

We exchanged pleasantries, and after a brief conversation about the direction of our path, he revealed something unusual: He knew about people more than anyone else, at least as I figured this out after talking to him. And he had learned this by picking mushrooms. This statement struck me, and our conversation quickly turned into a fascinating dialogue.

"So, you want to know what lies behind the reality we call life?" Martin nodded briefly and continued, pointing to the filled bag.

"You see, when you look at a mushroom growing in the forest, it appears inconspicuous", he said. "But the reality, which you may not even suspect, lies elsewhere. Mushrooms are the dominant form of life. In fact, the largest living organisms on Earth are mushrooms. The mushroom above the ground is just a fruiting body, or flower, intended for reproduction. What you usually do not see is the mycelium, or the fungal network — an enormous network of fine cells that permeates the soil and connects all aboveground fungal bodies everywhere".

"Right now, you're likely walking on one of these enormous mushrooms, even without realizing it," he nodded.

I quickly glanced at my boots with the decayed leaves stuck to them and pondered.

My companion continued: "A human, like a mushroom, is the fruit of consciousness, existing beyond the material limitations of space and time. It is connected to a vast network of pure information that links us all, and it's impossible to see in this material world. We all are manifestations of a vast network of spiritual information. Some call it 'God', others 'nirvana'. People have devised many names depending on their cultural characteristics. But the point is, it's just different interpretations of the same concept. This could be called 'cultural bias'. There are many types of mushrooms. But the essence is the same."

He nodded again at his bag of mushrooms and smiled. I was intrigued: "How can you prove this? Is this even a fitting analogy for human life?" I asked.

Martin replied: "Science proves it. But when too many possible conjectures converge and speak of the same thing, such coincidences rarely happen by chance. Ask, and I will try to answer, especially since we have enough time until we reach the hotel."

The meaning of life

This question was too obvious not to ask immediately, but I asked it after some hesitation, thinking that his answer would instantly clarify the absurdity of the analogy of people with the fruits of mycelium.

"Well, then, what is the meaning of life?" I asked.

He answered almost without hesitation: "People, like any living beings, appear here to gather information about the world, gain experience, and return it all back to the original reality — the place from where we all came. We are travelers, nourishing that other reality with new experience gathered in this world, emotions, impressions, and knowledge. Mushrooms are subjected to the influence of the sun, wind, and rain. They are made to withstand many external influences and phenomena. Just like the above ground bodies of mushrooms, our bodies are made of matter that is best suited for this world. We are mediators, connecting both realities".

He continued after a short pause: "Mushrooms, like people producing offspring, also give bodies to new life through their spores, or material substance."

Perhaps here I saw some analogy. But I was still not sure where he was leading his train of thought.

Anticipating my next question, he answered: "The purpose of life is in nourishing the entire vast mycelium with information. Without this information, the primary reality is just an empty vessel".

Then, after a slight hesitation, Martin clarified. "All life rests on an invisible ocean — a vast information pool. The more complex a life form, the stronger its ability to gather information and upload it into the reality of pure ideas. But animals cannot produce and analyze information about the world to the extent that humans are capable, as animals do not possess the ability for intelligent observation and analysis. Instead, they collect impressions about this word using conscience, since their intelligence is suppressed, and they heavily rely on hard-wired instincts to survive. Humans gather information without even thinking about it. Everyday activity, movement in space, interaction with living and non-living objects — all correspond to this purpose. The richer the experience, the more material for nourishing the original reality. Science, art, and culture are among the most important directions of information accumulation".

I tried to make sure I wasn't hearing things. "So, in your opinion, humans are just gatherers. Like mushroom pickers? And what do we need this for?"

"The question is not quite correct. We were given life. More precisely, the one who gave us life, that is us," he quickly parried.

Our brain

We started climbing the hill, and he continued. "Our brain, like the brains of other animals, evolved to gather information about the world around us. It has two goals: the first is the logical part with instincts for navigating space and collecting information about our physical reality. The second goal is to create a bridge for uploading and downloading this information into the world of ideas."

After a short pause, he added: "When you're awake, the interaction with the other information field is minimal. Your brain is fully tuned into the current reality. After all, this is its primary working environment. However, when you sleep, the logical centers responsible for functions in this world are inactive—the brain has a completely different function during sleep."

This time I tried not to interrupt and listened to the end.

"As is known, sleep is the most urgent necessity", he said. "More important than food. People can go insane or die if they do not sleep for several days. If sleep is more important than food, then it means that it is a key ingredient of our existence. On average, we sleep 25 years of our life. That's quite a lot! Even today, the fact that we do not know why we sleep perplexes science. You see, there are many theories about why we sleep. The problem of sleep is a question that has no precise answer. We do not understand how this physiological state that is experienced by the overwhelming majority of living organisms originated in nature. This feature does not make sense within the confines of Darwin's Theory of Evolution. Animals without the need for deep sleep should have a huge advantage over beings that spend half their time in limited interaction with the environment. The sleep process seems very disadvantageous for the organism, as it makes the animal extremely vulnerable to predators. Whatever sleep does, it must be worth the risk of turning off the brain. But then why, after hundreds of thousands of years of evolution, do we still not observe beings that do not sleep at all, but rest in some other safe way?"

Martin continued his monologue and said something so significantly important that it made me stop in the middle of the road: "People and any animals are collectors of information of this world," he explained. "This information needs to be synchronized with the network of the unconscious. Without this synchronization, the vast network of the unconscious, connecting the Universe, is empty. You need to sleep to upload the collected information into the world of ideas and meanings. Like any computer, frequent data backup to an external hard drive is desirable. A malfunction of our body is the most common cause of data loss in this realm. If the body fails before you share the data, there will be no chance for recovery. That's why it's important to regularly back up this information to prevent its loss."

The computer analogy was interesting. I pondered. Although at night I did not make backups — I just turned off my laptop.

"Exactly!" He exclaimed convincingly. "The data you collect, the visual and emotional representation of this world, you 'upload' every night into the reality I called `mycelium of meanings'. You cannot live without sleep for more than a few days. Your memory cannot keep the accumulated emotions and experience for too long. The human brain is too limited to accumulate such a large volume of information. Sleep also erases unnecessary information, minor details, and frees up space for new experiences. That's why we are unable to reproduce past details precisely. We only remember some fragments of the past and the most memorable events. The brain does more work than you think."

I started to understand what he was saying, but decided to clarify: "So, the newer experiences you have during the day, the more sleep you need, since uploading data takes more time?"

His answer made me realize that his work, apparently, was related to computers. "People who lead an active lifestyle need more sleep", he said. "Some may say — you need more time to sleep to restore the body's strength. But it can be said differently — you need more time for data transmission. When transferring information, the brain needs to be shut down. As you know, when you install patches for the computer, you need to stop the entire operating system. The operating system is in offline mode, but the computer is not — it does a lot of things but uses a different type of software that can work even when the regular operating system is in offline mode. In fact, at night the brain is active, but works differently. The brain does not rest when you sleep, that's for sure."

Martin continued, "The bridge between the brain and the mycelium of meanings is closed during the day. This is done to concentrate on this physical reality and avoid any external interference."

I noticed some discontent in his voice but didn't quite understand the reason.

"However, there are some deviations," he said. "Some people with an anomalous perception of the world may have very good access to the mycelium of meanings, even when not asleep. This can be somewhat exhausting. Meditations and prayers can also briefly open this bridge, but the connection is never strong and stable."

About dreams

The sun was almost set, casting an orange shadow over the pines. We had to move faster.

He continued. "During sleep, people connect to the mycelium of meanings to upload data. Since the information gates are open at this time, you can catch some echoes from another world. They are always superimposed on your experience and feelings. Such a symbiosis of information is interpreted by us as dreams".

"If dreams are fragments of our daily lives, I don't understand why they feel so extraordinary. In fact, most of what I see when I sleep is complete nonsense," I noted.

My companion smirked and picked up the pace: "Well, let's say, dreaming is a side effect, a feedback loop, distorted by your emotions. You can't control what you see in a dream, as your body and the logical and intellectual part of the brain responsible for existing in this reality are disconnected during data transmission".

It was too technical for me. Probably, my face reflected that. There was an awkward pause, during which I felt that his face became even more familiar to me, but somewhat different from a human's face when we started the conversation. Although now this dialogue had turned into something akin to an interview or a monologue. I couldn't keep up with his thoughts. And mine.

It seemed to me that he read my state from the expression on my face and continued more slowly, "Alright, let's use less technical terms. Do you see that dam at the other end of this valley?" He waved his hand.

I followed the direction of his hand and noticed the silvery glint of a small lake on the other side of the hill.

He looked at me and said: "This lake collects the melting water from that mountain. The dam at the base of the mountain opens when the water level reaches a certain point, and the water starts to flow, as the gates are open. The dam is closed now. If you are on the side where the water is collected, you see nothing behind the dates. These gates are made of solid steel and are opaque. This is analogous to your waking state".

Then he continued: "But when the gates are open, and the water begins to flow down the valley, you will see where it goes. Being in the upper lake, even if you swim under the water, you will begin to see the lower lake through the transparent water flowing down through the open dam gates. However, what you see will be distorted by the refraction of light through the flowing water. You will receive only a small and distorted fragment of information about the place where the water flows. For you, these will be deformed images interpreted by your brain as dreams".

For me, it began to make some sense. I tried to clarify. "Then, what do nightmares mean? Are they intermittent images of hell within the reality of ideas?"

"Oh no," he replied. "Hell doesn't exist. Nightmares are related to various physiological disorders of the body. You catch glimpses of new information, but distortions cause various illusions. Nightmares are like distorted mirrors, disfiguring information from another reality beyond this world. Your brain, receiving anxious signals from your body, introduces distortions and interprets glimpses of a parallel world as nightmarish visions".

After a small pause, Martin added: "But I don't think it bothers many. People see hundreds of thousands of dreams in their lifetime. Upon waking up, your brain cares about erasing them and tuning the intellect to manage your body in the matter and time of this world. In most cases, the information received during dreams is not what you need to fulfill your life's purpose here. The tiny number of dreams you remember are some deviations related to signals reflecting problems with the health of your body".

Information

Though I didn't ask anything, he continued to talk passionately, as if trying to clarify something for himself.

Summarizing, he added: "Information is what defines people's lives and why we are here. We are collectors of life experience. And carriers of it into the mycelium of meanings. Kazimir Malevich's famous painting "Black Square" is famous because it reminds us of this. In itself, it's worthless. This painting is worth exactly as much as the canvas and paint used to create this piece of art. It's the information associated with this painting that determines its value. The sky-high price that people attribute to it is merely a reflection of the created informational halo", he smiled. "For people, this painting, like many others, is simply a living reminder that information has value. Not the objects themselves. This brand-name leather bag (he waved his portfolio) is not the most functional thing – there are better bags out there. But you buy it because marketing programs our brain in such a way that certain items are priced at figures not corresponding to their real value."

"And what kind of information do we send... There?" I asked.

"About everything you experience here", he said and then clarified: "About what you see around you now. How you interact with other people, any of your experiences and feelings. Beauty! You, I see, are a scientist? You bring information about the details of the world's design. The thing is, to create this Universe, there's no need for the design and creation of every little part of it. It's enough to set the initial conditions and laws. And give the first impulse to start everything. And voilà! All the details of this world will automatically appear one after another, like popcorn from a popcorn machine. Heat the oil and open the lid — the kernels will start popping in all directions! You are precisely the one who gains knowledge about these sorts of details of this reality. This is valuable information. Moreover, people use their knowledge to change the original world. They complicate it, create previously unknown beauty and new environments for interaction, enrich their life experience, and initiate new trials for the soul. After all, a simple life in nature is not always interesting due to its monotony".

After death

I didn't remember why he switched to the topic of death. I didn't ask him about it, but Martin, as if it was nothing out of the ordinary, became passionately engaged in discussing this topic.

"We return to the mycelium of meanings", he said. "It contains all your experience and the experience of all other people. You can be wherever you want, using the memories, emotions, and information you've gathered in this world, as well as the information from billions of living beings from many worlds. You can build a house identical to the one you had on this planet. And meet the people you love".

"Like in a dream?" I interrupted.

This was his answer: "Yes, you can see deceased people in dreams, but you don't have full communication with them. In dreams, information moves in one direction — from you to the mycelium of meanings. You can only catch distorted glimpses from the world where the information flows. After death, the primary reality will be much more real than your world now. And communication with those you love will be more complete than in this world. You've already realized how much was left unsaid between you and your loved ones who have already left."

I was taken aback, and my heart ached. "Do you believe in what people see during clinical death?" I asked.

"It's similar to an altered dream. The part of the brain responsible for orientation in this reality becomes non-functional. The gates to the mycelium of meanings are wide open for the final upload of life information before the end of existence in this world. Communication needs to be efficient, as unlike in dreams, you don't have much time.

The gates are open much wider than during sleep. And so, you see much more beyond," he answered.

"Are there zombies?" With this idiotic question, I quickly tried to change the subject.

His response was so quick, it seemed like he knew the answer even before I asked about it. "It's like a mushroom that you just kicked. The place of the mushroom where it connects to the mycelium in the ground is the weakest spot."

To illustrate his point, he kicked a small yellow mushroom. "This mushroom is still good for a while, especially if you plant it somewhere. But this connection to the rest of the mushroom, giving meaning to the existence of its mycelium flower, is lost."

His answer was surprisingly clear. Now my thoughts were somewhere else.

Extraterrestrial life

We entered a roadside bar and continued our conversation. I was always tormented by the question: Is there life beyond Earth? I was interested in hearing his perspective, and I was not disappointed in my expectations — when he began to speak, it seemed to me that he was just waiting for my question.

"In this world, you will never find very advanced civilizations. As soon as a civilization reaches a high level of technological development, it leaves this world and heads to the mycelium of meanings, where it remains. New spheres of life are created where the resource expenditure is smaller than in the world made of matter."

Seeing the confusion on my face, Martin explained. "The main goal of any civilization is to find the open gates to the mycelium of meanings, where members of the civilization can communicate with their equals. Developing technologies that allow traveling and overcoming many light years to find similar beings is not an optimal solution. The limitations of the speed of light and the vast expanse of space make interstellar flights impossible."

I am only now beginning to understand that he is referring to the Fermi paradox – the contradiction between the lack of evidence and high estimates of the probability of the existence of extraterrestrial civilizations.

Religion

I'm not very religious. Perhaps I could describe myself as a spiritually minded person. But still, I wanted to know what Martin thinks about religion.

"Different religions are different reflections of the mycelium of meanings. All human superstitions and all human experiences are a direct result of rare moments of interaction with another reality. Each culture receives knowledge about the primary reality at different historical times. Therefore, different religions differ in detail, but the spiritual essence remains unchanged," he replied.

I had nothing to add here. We placed our empty cups on the table and, having paid, and got up from the table.

Inequality

We left the bar and headed towards the hotel complex, visible in the distance. It was getting dark. The dull sound of a snow avalanche falling high in the mountains carried through the air. My attention was focused on the rocky mountain path, along the edges of which mushrooms of various sizes and types grew.

I decided to change the subject of our conversation. "Okay. But why is there inequality among people?"

After a pause, he continued, "People are different, like the mushrooms under our feet. Some are big, others are small. It's all because of the randomness of this material reality. People are prisoners of randomness and circumstances. Happenings of things in the material world are described by a bell-shaped probability function, which has large tails for extraordinary events of small probabilities. Mushroom sizes are not uniform, even if they come from the same evenly distributed mycelium. The difference in the conditions in which they are born affects their appearance. Some appear on depleted soil, others under too bright sun, others just in the wrong place under a rock. And some, not having time to grow, perish because they started to grow too close to a busy road."

"Yes. I see this randomness when some people are born into wealthy families, and others in low-income families. Wealthy children become wealthy adults, and poor children become poor adults," I noted. "What about those who became rich on their own, like Steve Jobs?"

My companion smiled, "Well, it's almost the same. This situation is not much different from being born into a wealthy family — it's also the result of randomness and luck. You see, there are many hardworking and creative people, but only a small portion of such people become as successful as Jobs. It's just the science of random numbers. Any bell-shaped probability distribution must have a minimum, an average, and a maximum. People who become rich without external help are at the very edge of the maximum in such distributions. It's related to luck and suitable conditions from a chain of random events. Someone has to be Jobs, someone has to be less fortunate."

"What about geniuses?" I interrupted.

"Geniuses do not exist. There are more and less hardworking people, there are lazy ones, there are people smarter than the average, there are those who simply have different priorities. Einsteins are everywhere. People can only fit a few names in their brain, not thousands of names of people who step by step move progress forward. They prepared and paved the way for discoveries. The concept of geniuses is the result of the human brain's limitation in understanding how progress evolves, which is an intellectual movement of a vast number of people."

Martin continued, "Did you hear the sound of the snow avalanche a few minutes ago? Any major scientific discovery is like an avalanche. One snowflake can cause an avalanche to descend. But before that, a lot of melting snow needs to accumulate on the mountain slope. The person making a scientific discovery is like a random snowflake that triggered the avalanche. Your brain can't remember or even comprehend the work of a huge number of people who influenced the development of scientific and technological progress. But it can remember a few names, such as Jobs or Einstein. And as soon as randomness creates a wealthy or relatively famous person, everything else becomes easier. As is well known, the more money you earn, the more money will come. The same is true for fame."

I too had thought about this but decided not to interrupt.

"The most successful people in society are just the luckiest people, singled out from a huge, hard-to-comprehend number of hardworking and talented people by a random sequence of events and trivial luck. Rich and poor... They are the same to the mycelium of meanings. But one thing should be remembered: It is impossible to achieve the goals of presence in this world by isolating oneself from its reality. If you are rich, the temptation to isolate yourself from the world is very great. Only a few wealthy people can overcome such temptations and achieve their true goals. Many disappear into their artificial world, isolated from other people. Too many of them attempt to maintain their lifestyle with their islands, castles, and huge homes with fences. Or to be constantly on the road."

I looked at my acquaintance and saw that he was trying to find the right words. He finally said, "For this reason, the wealthy are rarely the most fortunate collectors of experience and emotions. Their world is too artificial and limited. Often, they are lonely. Their attachments boil down to things, not people. They cannot casually communicate with ordinary people and have friends among them. This is not exactly the experience that is needed. Without going through difficulties, it is unthinkable to enrich the soul and mind, it is impossible to comprehend love."

I remembered a long-forgotten phrase from the Bible: "It is easier for a camel to go through the eye of a needle than for a rich man to enter the kingdom of God."

He smiled and picked up, "In general, Earth is a damn hard place. You never know where you'll find or lose something. Everything is so complicated. But the lessons we learn here are quick. Quicker than anywhere else... At least, most of us do."

Justice

Is there justice in this world? This question has troubled me for quite some time. I continued: "People are constantly being killed. Why was fascism allowed to annihilate millions in pursuit of ideological agendas and material wealth!? And what does it mean to incinerate children and women with a nuclear bomb to save the lives of soldiers whose profession is to die in battles?"

Martin's answer astonished me, "Murder does not significantly affect the mycelium of souls itself. When you pick mushrooms, their bodies are cut off at the point of connection with the ground, but the mycelium in the soil is not damaged. Murder does not harm the soul of the victim. The person who becomes a victim loses very little — only their body. If they wish, they can reappear in this reality at any place and even time."

After a pause, he continued: "However, crimes damage the 'mycelium' of the soul of the person who committed such acts. This information is transmitted to the mycelium of meanings during their sleep. After their death, the criminals will find themselves in the reality they brought there. It will be the pain of their victims. They will not find a way to build comfort in the new world. The only choice for them will be to return to this reality again and make changes in their life that will make existence within the mycelium more peaceful and happier".

My companion smiled and looked at me, "So, in this sense, justice does exist. The perpetrator needs to return to the material world for as long as necessary to find peace, love, and happiness in this world and purify their soul. To gather such positive information, to do good, and to bring this good where they can experience more consolation after their death. Kindness and love are the main assets we all chase here to make our existence within the mycelium of meanings more comfortable".

"But why do all these criminals appear in the first place?" I asked.

"For various reasons... A person, born in this reality, forgets why they are here. This is what free will is — we appear here, but without much guidance from the other side. We are given freedom, without the 'baggage' of things that are not needed here. It's hard to make a choice between various possible actions. Your soul or spirit, a small drop of meaning, is completely autonomous in this world." And then, — he noted, — "Even mushrooms have worms. They can affect your actions because they eat away your connection to the other world."

I didn't ask for clarification on the analogy with worms...

Love

The landscape strangely transformed, and I began to feel that the ground had become unusually springy. There was a smell of dampness. I realized we had come to a swamp, or some dried-up lake. The ground, covered with moss, quivered. It was as if we were walking on a stretched trampoline. There were no forests in sight, only small bushes alternated with rotting birches. The bushes next to me rustled with each step I took. It looked almost unreal.

I noticed that the face of my companion also expressed surprise, but he continued, "We are travelers here. Our findings are what hold value for us. In this world, we must gather happiness, love, and carry emotions, experience, and everything that will fill this world with meaning and happiness. There, we use it as building material to create a happier existence."

I picked up, "If you loved in this world, you would carry that feeling into another reality and will be able to reunite with those you loved."

Martin continued: "But love can also be brought from the other side, through an illusory connection that can be opened at night. Love at first sight, as it's commonly called, is usually a distant echo, a shadow of love brought from a parallel reality. This love can also be built on the basis of previous experience, taken from other similar realms and crystallized inside the mycelium of meanings long before your arrival in this world".

The Question of Origin

"Where is this mycelium of information of yours? As I understand, it's not located in our Universe? But how can it be in contact with our brain?" I asked.

He glanced at me quickly: "The mycelium isn't here. It doesn't have a spatio-temporal form. But you can definitely find its traces. Lifting a mushroom, you'll find the mycelium, but you need a tool."

And then his explanation astonished me: "The one who created this world can only be in one place. For our world — that's in the distant past. The very moment of the formation of this Universe — the Big Bang — is an illusion that looks like the past to you. In reality, for another world, filled with meaning and love, there is no time. It is, was, and always will be."

Burden of Proof

Finally, we emerged onto drier ground. The forest with its beaten path loomed ahead. I sighed with relief. Walking across the quivering mossy tussocks was not part of my plans.

"Well then, how do you prove all this?" I asked.

Martin replied with a smile: "Well, from what I've told you, you can start to guess. Here's a simple experiment: If you've had a good night's sleep, you won't want to sleep again until the evening. Your brain is fresh and ready for a new day. You've shared your experience and disconnected from the mycelium of meanings. But accidents can happen even in the morning. Life-threatening situations trigger a new near-death experience, during which information will be synchronized. But the time needed for this synchronization is much less for people who have just woken up. So, a near-death experience in such cases is significantly weaker, compared to cases where death occurs in the evening for people who have not slept for a long time."

Looking at my astonished face, Martin added, "As you understand, this experiment is complex, but real. Have you ever had dreams that are impossible to explain or dreams that defy any logical explanation? Have you dreamt of melodies you've never heard before? Or people and places you've never seen in your life, but in your dreams, they are familiar down to the smallest detail? Have you ever touched something bigger, happier, full of joy and relief in your dreams? If you don't believe that the miracle of dreaming can be simply a byproduct of your resting body, then the only explanation lies in one thing: dreams are echoes of an incredible reality beyond the material world."

My interlocutor paused. It seemed to me that his silhouette in the pale light had started dissolving into the evening air.

"All you need to know is within you. Just ask the question…". These were his last words I could hear.

I woke up in a cold sweat and looked around. I felt bewildered and completely disoriented, as often happens when you're woken up in the middle of the night. Soon my thoughts began to clear. I was sitting on a bus approaching my stop. It was a regular day in Chicago. The noise of cars was heard outside. I looked around, mechanically picked up my new leather briefcase filled with my lectures and stepped out onto the bustling street. The morning cool mist began dissolving the last traces of my dream. A new day had begun.

13 The origin

The question of the origin of the Universe is a question of science, faith and philosophy. In response to this question, there are two camps: those who believe that the Universe formed by itself (or has always existed), and those who believe that it was created.

13.1 Universe from nothing

Until the 20th century, most people believed that the Universe had neither a beginning nor an end. The Universe was boundless and had always existed. It was the prime cause of biological life, consciousness and information. In the 20th century, this paradigm shifted to another. Most scientists began to assume that time and matter emerged from a singularity as a result of the Big Bang. Where it came from is unknown. But perhaps one day it will be understood.

A naturalistic (also known as scientific) explanation might be as follows: Perhaps, there are an infinite number of universes. Our Universe is the way it is because we live in it. Before our world appeared, there was some other Universe that gave birth to our Universe. For example, it is conceivable that it emerged from some quantum fluctuation — from nothing (or something not yet understood). Of course, to say that a fluctuation can arise from nothing is a philosophical contradiction. Perhaps there was a certain medium that existed outside our Universe. Let's assume there are some processes that create an infinite number of universes. They are born like soap bubbles. The unburst bubble-universes survive. It is these that have a fine-tuning of physical parameters to exist for a long time, which was sufficient for the spontaneous origin of the first cell, multicellular organisms and their evolution into complex animals and humans. And having achieved significant progress in understanding the world, people came to the conclusion that all the parameters of their Universe were amazingly tuned for their own existence.

In this paradigm, life is merely reproducing imprints of clusters of molecules as a result of the interaction between various chemical and physical processes. When first primitive life emerged through random collisions of atoms and molecules, everything else is the result of evolution, according to the natural mechanisms of the Theory of Evolution (see chapter 3.6 Theory of evolution). All the information that surrounds us came about on its own, as soon as the simplest biological life appeared. Biological information is created by matter —

simple as that. The evolving life itself is the producer of new information. And then this information grows geometrically, creating complex organisms and humans. Billions of years can do anything, we just have to wait. Humans have no purpose and no meaning other than what they invent for themselves, i.e. it is a pure subjective notion. Laws of nature are also subjective concepts in people's heads. They describe how one material phenomenon depends on another. The laws themselves do not have any objective meaning. Like everything else around us.

Recently, quite a few books have appeared where science is used to explain why and how the Big Bang occurred. For example, theoretical physicist and cosmologist Lawrence Krauss (Krauss 2013) argues in his book "A Universe from Nothing: Why There Is Something Rather than Nothing" how the Universe came into existence from nothing. However, he redefined "nothing" into "something" without the reader noticing. This "something" is just a non-classical definition of space-time. It contained some "stuff" and the necessary quantum effects, together with some embedded laws, to create the Universe.

In the book "The Grand Design" (Hawking and Mlodinow 2010), theoretical physicist Stephen Hawking, co-authored with American physicist Leonard Mlodinow, writes:

"Because there is such a law as gravity, the Universe can and will create itself from nothing. Spontaneous creation is the reason there is something rather than nothing. ... There is no need to invoke God."

This statement attempts to answer the profound question of the origin using terms that extend far beyond the scientific method of knowledge. It undoubtedly contains an element of faith — that gravity alone can create the Universe from nothing, although there is no slightest reason to think so. But most importantly, it assumes that gravity and the laws of physics already existed before the Universe appeared. That is, it again presupposes that "something" was indeed always there.

It can be agreed that this line of reasoning is a fundamental logical fallacy (Lennox 2021). It's incorrect to assert that something which does not exist, can create itself. The laws of physics, such as those describing gravity, cannot precede the existence of the Universe. What requires creation cannot form itself. Hawking writes that "philosophy is dead" because philosophers have not kept up with the latest achievements in physics. However, he makes elementary mistakes in logic, which are unacceptable for philosophers. For example, the concept of God as a "god of the gaps" in knowledge is no longer relevant as a contemporary theological concept. Today the argument that God is a way for "filling gaps" in our understanding is not taken seriously among contemporary philosophers and theosophists. Indeed, aborigines might have explained some natural phenomena, such as thunder, using the analogy of a war of superhumanly powerful beings, but the concept of God underwent a radical transformation in the Middle Ages.

The concept of God evolved specifically in connection with attempts to find answers to very profound questions. It was developed and supplemented in the Middle Ages precisely because of the understanding of eternal questions that needed to be asked, not because of ignorance or the lack of knowledge in detailed mechanisms of nature. Even at that time, ignorance about the origins of natural phenomena, such as wind, lightning, or thunder, was not something that required a God. The ability to separate the question of the technical design of something from questions of origin and meaning is a sign of a high intellectual level and attempts to delve into the essence of the reasons for existence as such. We will return to this question in chapter 18 Hypothesis of God.

Another book, "How It Began" (Impey 2012), doesn't even attempt to answer the question posed in its title, as everything described only pertains to events that occurred after the Big Bang created space and time.

I don't criticize materialists advocating the concept of "everything from nothing" at all. To justify such a materialistic approach, an incredibly vast imagination and the ability to create logical (but, usually, unprovable) constructions from scientific categories are required. By and large, these researchers are just as much dreamers, in the best sense of the word, as those who argue for the presence of a designer who launched the entire project called the "Universe".

Finally, not answering the question about the origin of the Universe is a common approach to the problem among intellectually inclined materialists. Not attempting to answer what we definitely don't know is often seen as a sign of seriousness and maturity in reasoning. However, as we have shown many times in this book, it is precisely the fantasizers and those who are not afraid to ask impossible questions and answer with improbable hypotheses constitute the most active part of the scientists advancing the understanding of the world.

It is highly probable that the true explanation of our world is much more unimaginable than what materialists and idealists propose (see chapter 3.9 Idealism and information). We simply lack the categories to describe the reality lying beyond the stage of the material world. If materialism strictly delineates the boundaries of its concepts, then those who admit that everything around us is created leave more room for incredible explanations.

13.2 Universe as creation

The concept that everything around us is a creation deserves no less attention. This is not an attempt to evade the question. It is an endeavor to use a different approach for the answer and to look into the root cause. If you believe that what you see are traces of information that leads to some complex functionality, you will start searching for its author. In this case, you must adopt a somewhat different approach in such searches, compared to those who think in terms of luck in random processes and natural laws originating from nowhere.

For instance, imagine you are in a desert on an uninhabited planet. You discover something quite complex, with numerous dashes and dots. Assuming it to be the creation of an ancient civilization, you might start excavations at this site. Your interpretation of the data prompts a specific way of seeking truth. However, if you are a materialist and do not accept the existence of a creation, you might instead study the air currents in this desert and the characteristics of soil erosion, trying to understand how natural processes led to such an orderly arrangement of signs. Which approach would lead you to the truth more quickly?

Thus, the answer to the question of the origin of the world can look like this: The Universe was created, along with space and time. Since space-time was also created, the one who created everything was outside space and time. Hence, this Being (or God) was not "made". He is the first cause. In this book, we will consistently refer to God using the pronouns "he" and "him". While acknowledging that God transcends biological gender, human comprehension often finds this concept challenging. Therefore, we opt for more traditional Christian terminology, albeit using lowercase for "him".

In this approach, the fine-tuning of the laws of nature was a rational product of creativity. Since the Universe was created, and its laws were fine-tuned for life, the one who created it made such a decision guided by some reason. Therefore, he imparted meaning to the Universe and was rational. He filled the Universe with information to create life and man, perhaps, in his own likeness of spirit.

The doctrine of the creation of the Universe presupposes that the laws of nature were designed before matter and energy were created. All processes began to change according to these laws using time. Humans can recreate such laws using abstract thinking. The Creator left his traces in this creation. They are entirely sufficient for those who feel that the world is much more wondrous than it appears. But such traces are not enough for those who can reduce the entire miracle of the Universe to random collisions of atoms and molecules. Such uncertainty leaves us with free will (see chapter 17 Free will).

For those who believe in the creation of the Universe, a system of belief emerges. For them, our world has a certain purpose. Since the Universe with its space-time was created, then the creator itself must be beyond space and time. Information is formed only by a mind that sees meaning in its creation. Thus, life, as a product of information, was created.

Does this approach cancel out science? No. Science is used to understand the structure and beauty of such a creation. One of the greatest scientists, Isaac Newton (1643 – 1727), believed that science without God is meaningless. He said, *"Gravity explains the motion of the planets, but it cannot explain who sets the planets in motion"*. We will return to this issue in chapter 18.3 Scientists at a crossroads.

311

13.3 Explanations and fantasies

If you take a closer look at these two versions — a Universe without a creator and a created Universe — it becomes clear that both explanations involve the concept of faith. It all depends on which reality you are willing to accept. There is nothing "scientific" about either of them because science relies on methods that cannot be used to test such scenarios.

The concept of a Universe without a creator contains a lot of speculation, while manipulating scientific terms. These belong to the category of scientific fantasies, where scientific concepts lend credibility and recognition within the scientific community. For example, a person with critical or scientific thinking is much more likely to believe in a Universe created by a quantum fluctuation, as the words "quantum" and "fluctuation" lend a certain scientific hue, acceptable among scientifically-minded researchers. The question of in what medium such a fluctuation happened, if there was nothing before the Universe existed, requires less justification.

Books that explain the origin of the Universe using matter itself (Hawking 1988) (Krauss 2013) contain a vast number of unscientific assumptions. Possible scenarios about the birth of the Universe, using the laws of physics, are as pseudoscientific as any other explanation, with the only difference being that such fantasies have "merged" with scientific concepts and terms, creating an impression of scientific legitimacy. It must be said that I am not at all against scientific fantasies or any creative approach to the question of origin. However, the criticism of such approaches (Lennox 2019), (Geisler and Turek 2004) (Meyer 2021) should not be perceived as an attack on the scientific approach to a question that has no scientific explanation.

If we assume the existence of an infinite number of universes that led to the creation of this one, where the laws of nature are fine-tuned for complex life, we quickly come to the conclusion that such an explanation is absurd. It involves an infinite number of unknowns to answer this (single!) unknown. Such a hypothesis explains nothing, as you need to answer the question of how this infinite number of universes came into being. Furthermore, life requires information. The creation of information by a random physical process is an even more fantastical hypothesis.

The assertion that science can explain what happened during the creation of the Universe is utterly implausible. Science primarily deals with describing phenomena or the structure of objects using observations and formulating hypotheses. Understanding this without experiments in a laboratory or isolated conditions is incredibly difficult. When dealing with historical information-rich events, extrapolating knowledge to phenomena that occurred in the distant past without any experimental confirmation requires many unprovable assumptions. Whether to believe in them or not goes beyond the scope of science.

I'll give a simple example. It has been already discussed before, but I'll repeat it again for those who did not read the beginning of this book. Here it goes: How do you know what you ate 3 days ago, if you don't remember? You also have no record because you don't keep a diary. Perhaps you have receipts from the store, and you could sift through the trash bin to see what's there. Even if you found eggshells in the trash, which are roughly three days old, you can only make an assumption based on what you generally like to eat and what you could make for dinner. You might even experiment in your kitchen with eggs and make a few dishes using potentially possible recipes. But that's where your "knowledge" ends. You can only assume the most probable — that you ate scrambled eggs because you like them. But you'll never prove it. The notion of "probable" remains insufficient as conclusive evidence within the realm of scientific inquiry. There are many scenarios where the eggshell ended up in the trash, but you didn't make scrambled eggs. Maybe you accidentally broke the eggs bringing them home from the store? Perhaps a friend visited and helped you make a new dish?

If you are unable to respond to even such a straightforward inquiry, how can you answer a question about how our Universe and life were created billions of years ago? Where do we know from that the laws of natural science, with all their constants, have not changed over time? Maybe the correct answer should be based on the entire body of human knowledge, including the exact sciences, philosophy, culture, religion, and of course, using historical records of the past?

13.4 Myths and religions

Most myths and religious traditions assert that the world was created according to some design. At the end of the 19th and into the 20th centuries, this was viewed as ignorance, as scientific and technological progress proved its success in understanding the world, and nothing beyond science was required to explain the Universe. As a result, the philosophical way of thinking took a back seat. If we can figure out the laws by which the world of atoms and molecules works, and we can create various contraptions, machines and computers, then what more is needed? Why not use this method to explain the formation of the world? As we have said, such a view is an attempt to construct myths using science. However, this is not something new, as at every stage of human development, the concepts at hand and their relevance at the time became the "building material" for fantasies trying to explain the appearance of the Universe.

However, one amendment needs to be made for the most ancient forms of culture and tradition. In virtually all myths, the creation of the world is carried out through the word, the pure thought of the creator, a dream, or some actions of a divine being. The creation of the world from nothing is found in the myths of ancient Egypt, cultures of Africa, Asia, Oceania, and North America. The myth that God created the world from nothing – ex nihilo – today holds a central place in Judaism, Christianity and Islam. In ancient times, the very idea that the world is infinite and exists forever (as was believed until recently), or that it could be created from nothing by some inanimate natural phenomenon, seemed like a ludicrous idea. Indeed, at that time, man did not know as many laws of nature and physics as we do now. Nonetheless, he was acutely aware that complex, organized things could only be created.

The fact that the foundation of our reality lies in ideas, rational design, and the primacy of the spiritual over the material is clearly traced in the Bible. In the Old Testament (Genesis, chapter 1) God speaks the word and creates our world:

1. *In the beginning God created heaven and earth.*
2. *Now the earth was a formless void, there was darkness over the deep, with a divine wind sweeping over the waters.*
3. *God said, 'Let there be light,' and there was light.*
4. *God saw that light was good, and God divided light from darkness.*
5. *God called light 'day', and darkness he called 'night'. Evening came and morning came: the first day.*
6. *God said, 'Let there be a vault through the middle of the waters to divide the waters in two.' And so it was.*
7. *God made the vault, and it divided the waters under the vault from the waters above the vault.*

8. *God called the vault 'heaven'. Evening came and morning came: the second day.*
9. *God said, 'Let the waters under heaven come together into a single mass, and let dry land appear.' And so it was.*
10. *God called the dry land 'earth' and the mass of waters 'seas', and God saw that it was good.*
11. *God said, 'Let the earth produce vegetation: seed-bearing plants, and fruit trees on earth, bearing fruit with their seed inside, each corresponding to its own species.' And so it was.*

The main point in these lines is not the sequence of the creation of light and water (which can be debated), but the very fact that creation occurs after "He said". This implies that a rational design, transformed into information, creates this world — both inanimate matter and life.

When this part of the Old Testament is criticized, all attention is given to the irrational order of the world's creation, and the use of the word "Day". This irrationality arises only because the creation process is interpreted exclusively literally. Critics of religion assume that by creating one part of the Universe (say, water as a chemical element), that part immediately becomes a "working part". Such a view is complete nonsense to every engineer or programmer. They well know that the process of creating complex machines or virtual worlds in a computer does not occur in a sequence where one created part is fully functional before all components are assembled together, especially when one part depends on another. To create something complex, where one part interacts with another, first, all these parts must be planned and created. The order of creating such parts can be quite arbitrary. And only at the end of the work are the created parts joined together. Secondly, this world was made beyond the limits (which is quite obvious) of this reality, which the word "day" belongs to ease human comprehension.

This example might clarify what I mean. Imagine you're creating a computer game in six Earth days. Initially, you wrote the program code for the sky and the Earth. Then you programmed light shadows, the surface of our planet, and so on. As a result, the entire simulation of the Universe, in its beauty and perfection, was fully formed over six literal days, but in that world, it might not be 24 hours! On the 6th day, all components of the game-simulation were completed, and the world came to life.

Perhaps in the first seconds of such a game, the Big Bang was played out, simply so that participants of the game at some point in its development wouldn't guess that the game had been turned on suddenly. Or the parameters of the game were selected in such a way that it looked as if there had been a Big Bang in the distant past — again, so that a materialistic description would become a direct alternative to a world created by design for those participating in the game.

Then the creator leaves a short note of a few lines about how he made this world, hiding it somewhere in a file inside the game. And on the 7th day, he went to rest. How could a character of such a game, like Moses, describe the structure of this game in his Old Testament, after reading this note? I think understanding what was written in that note was difficult. Moses attempted to describe the creation of his world using concepts that are accessible to characters of such a game.

There can be many fantasies, but the most important thing here is not the sequence of the created parts of this world or the days. The principle that matter is secondary and created from a rational design is the main theme of the Old Testament.

In my view, the Old Testament (Genesis) can easily be interpreted in a non-contradictory manner:

- The world was created outside this Universe (which is logical), and the word "day" reflects the period of time of that reality where it was made. It could have indeed been their day.

- The Universe was created in parts due to its incredible complexity. Once its parts were created, they were put together. The "discrepancy" in the sequence of the world's creation laid out in the Bible, as critics think, is simply used to explain that the world was created like an intricate machine. For example, like a computer or a car, when individual parts are created in some arbitrary sequence at the initial stage of creation. This simply points to the creative process of a complex system.

- The world was created according to blueprints or a plan, as it arose in the creator's consciousness and initially existed as information. The Bible insistently emphasizes "He said" before making each part.

The clearest thought that the basis of matter is idea and information is written in the Gospel of John. We read (John 1:1-18):

The Word Became Flesh:

1. In the beginning was the Word, and the Word was with God, and the Word was God.

2. He was with God in the beginning.

3. Through him all things were made; without him nothing was made that has been made.

4. In him was life, and that life was the light of all mankind.

...

14. The Word became flesh and made his dwelling among us. We have seen his glory, the glory of the one and only Son, who came from the Father, full of grace and truth.

The text clearly states that at the very beginning of the Universe's creation, there was the Word, meaning an element of information carrying a certain meaning. This Word was "God Himself". Everything about the Universe started thanks to the God-Word. Why is the term "word" used? And how else could it have been written at that time for people to understand the meaning of such a text? How else could a person living about 1900 years ago describe the beginning, where information and intelligent design is the source of this world?

Further, the text says this information was about the life of people, and links it with "light". Perhaps at this moment, the text speaks of how exactly information defines life. As we will see later, light is often associated with something immaterial, yet at the same time, containing meaningful information. Perhaps at this moment, the text speaks of some universal intelligence, which is not material in our understanding. Clearly, humans in their biological shell did not yet exist.

Then, the information was used to create flesh, that is, a biological shell for life in the material world of this Universe. Interpreting this part of the Bible in such a way becomes quite understandable and is fully consistent with the immaterial foundation of this world — the concepts of meaning that can exist by themselves.

Critics of the religious explanation of the Universe will be right in one thing — religion uses dogma to describe the origin of the world. It will never change. On the contrary, science uses facts. This means that scientific hypotheses and models of world creation can change over time, getting closer to the truth. This is a significant difference from the Bible. Here, it should be noted that science can indeed change its views on the description of the world. It's foolish to deny this. But the fact that a certain mechanism or phenomenon resulted from other processes or phenomena is also a dogma for science. Religion does not claim to explain the details of how physical or chemical laws work.

14 Light

I have long pondered whether to include a discussion of the concept of light in this book. What, after all, does light have to do with information? We know that light is simply electromagnetic radiation that our eyes perceive (within a rather narrow frequency range). The carriers of such radiation are photons — quanta of light. Photons have no mass and they "race" at the highest speed — the speed of electromagnetic waves in a vacuum, which is approximately equal to 3×10^8 meters per second (m/s). In some mysterious way, this speed enters into the fine-structure constant, which determines the strength of the interaction of photons with matter and compensates for incredibly small quantities, as we have already mentioned in chapter 9.7 The fine structure of the world.

Here, we could conclude this chapter. Yet, why not venture into a bit of speculation?

14.1 Little bit of science

But first, let's talk a bit about basic physics. An electromagnetic wave is characterized by a parameter — the number of crests that pass by an observer in one second. The frequency, denoted by the Latin letter v, can be used to determine the wavelength $\lambda = c/v$. The energy quantum is related to the wave frequency as $E = h \times v$, where h is Planck's constant, which is discussed in chapter 8.6 Again about information.

The human eye perceives electromagnetic waves with wavelengths of 380 – 780 nm (nanometers) as light. One nanometer equals one-billionth of a meter. Electromagnetic waves with a low frequency correspond to radio waves with a wavelength of 0.01 cm (centimeters) and longer. Electromagnetic waves with a lower frequency are gamma rays (0.01 nm and shorter). The shorter the wavelength and the higher its frequency, the greater the photon's energy. See Figure. 14.1.

At the Large Hadron Collider — the most powerful particle accelerator in the world, located at CERN in Switzerland, I led a team of scientists who studied electromagnetic radiation with a wavelength of 1.24×10^{-9} nm. This corresponds to 1000 gigaelectronvolts or 1000 billion electronvolts. Such high energies for quanta of light were obtained by humans for the first time in an experimental setup. These photons were transformed into other particles that could be observed in the detector.

In the microworld, photons are capable of transforming themselves into other elementary particles with mass. For example, matter can be created from two photons. The first published calculations of electron-positron pair formation in photon collisions were performed by the Soviet theoretical physicist Lev Landau (1908 – 1968) in 1934. To create much more massive pairs of particles, such as a proton and antiproton, photons with an energy of more than 1.88 gigaelectronvolts are required. The higher the energy of the photon, the more matter can be created.

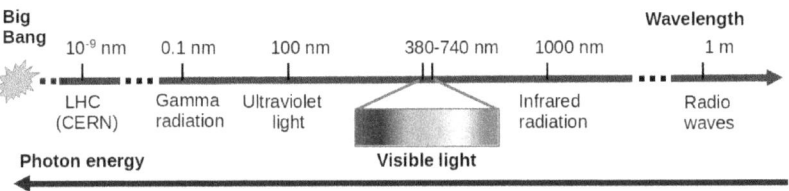

Figure. 14.1. *The spectrum of electromagnetic radiation, arranged in increasing wavelength (or decreasing photon energy). The range of visible light is marked by the interval 380 – 740 nm.*

David Bohn (1917 – 1992), one of the most significant theoretical quantum physicists of the 20th century, had a special role for light. For him, "matter is condensed or frozen light". Here is some of his thoughts (Weber 1990):

".. All matter is a condensation of light into patterns moving back and forth at average speeds which are less than the speed of light. Even Einstein had some hint of that idea. You could say that when we come to light, we are coming to the fundamental activity in which existence has its ground, or at least coming close to it. ... As move faster and faster according to relativity your time rates slow down and the distance gets smaller, so as you approach very high speeds your own internal time and distance becomes less, and therefore if you were at the speed of light you could reach from one end of the Universe to the

other without changing your age at all. ... We're saying that existentially speaking or logically speaking, time originates out of the timeless. "

Some of you probably know that a photon traveling at the speed of light does not "feel" either time or space. For photons, the observable Universe turns into a plane of infinite thinness. And time simply ceases to exist, while the cause and effect of events start to occur simultaneously. Essentially, for a photon, you have neither been born yet nor have long been dead. From the perspective of this particle, everything around is in a kind of imaginary world where all processes happen at the same time.

From the point of view of complex biological organisms, light is something special. One way to understand the connection between light and information is illustrated by the simple fact: When you open your eyes, you see light, and as a result, you begin to receive information about the surrounding world. Photons of light are absorbed, reflected, and emitted by various materials that make up physical objects. It can be said that light is a necessary condition for the human brain to receive most visual information. It has been calculated that about 80% of all information we receive from the surrounding world is thanks to light. Researchers have calculated (Balasubramanian 2006) that the human retina can transmit data at a rate of about 10 million bits per second. The retina of the eye is a part of the brain that has grown into the eye. For comparison, Ethernet (a family of data transmission technologies between devices) can transmit information between computers at a speed of 10 to 100 million bits per second. Although the human brain receives information using photons, for photons, you simply do not exist in the same sense as we understand "existence".

14.2 Little bit of religion

The theme of light permeates all Christian doctrine. In the Old Testament, the very first lines say:

"God said, 'Let there be light,' and there was light.

God saw that light was good, and God divided light from darkness. "

Later, according to the Old Testament, the luminaries themselves and the Sun were created. If this text truly is a historical record of how the world was created, then chapter 13.4 Myths and religions explains that such an order in this text can make sense if one assumes that the design was not sequential, as is the case in reality when creating complex structures or systems. If our Universe is considered as such.

Continuing to read the Gospel of John, we find that information is identified with light and life. We read (John 1:1-18):

"1. In the beginning was the Word, and the Word was with God, and the Word was God.

2. He was with God in the beginning.

3. Through him all things were made; without him nothing was made that has been made.

4. In him was life, and that life was the light of all mankind.

5. The light shines in the darkness, and the darkness has not overcome it.

6. There was a man sent from God whose name was John.

7. He came as a witness to testify concerning that light, so that through him all might believe.

8. He himself was not the light; he came only as a witness to the light.

9. The true light that gives light to everyone was coming into the world."

According to these lines, the light mentioned here has nothing to do with visible radiation. It's some other kind of light. The Russian language has long absorbed this distinction. When someone dies, people say — "He has gone to *that* light" or — "She is in *that* light". Religion suggests that there is some other light to which people go. Such light is often seen by people during clinical death, observing a light at the end of a tunnel (chapter 15.3 Consciousness outside the body).

In Christianity, the issue of light is resolved by distinguishing between "created" and "uncreated" light. The first type of light is perceived by our eyes. This is visible light. It is considered created light because it comes from the sun and stars.

Uncreated light refers to Divine light. This is the light that John speaks about. It is into this light that people go at the time of death, uniting with God. Since there is no physical body, there is no access to created light, which exists exclusively in the material world and interacts with our biological shells.

From a religious point of view, both types of light are a necessary condition for existence. In this world, we use created light to receive information about the world. But as soon as we transition from this world to another, we connect with the original uncreated Light (capitalized) which is God or the Word.

14.3 Little bit of fantasy

As someone once said, physicists are bad philosophers. But there is no physics without imagination. Let's try to generalize everything we know about light (or electromagnetic waves) and fantasize:

- Photons are the fastest carrier of information we know. Nothing can travel faster than photons.
- Photons with high energies can create matter. For matter born from photons, the moment of the matter's birth will always be in the past. But not for the photons that gave birth to the matter.
- If you fly in a bundle of photons (if this is possible!), then for you, all space will "collapse" into a single plane. And time will cease to exist. Cause and effect will happen simultaneously.
- There are many discoveries in this world precisely because of electromagnetic radiation (see chapter 15 Knowledge through eclipses).
- Light is the central concept in the creation of the world, according to the Bible.

As we discussed earlier in chapter 9.7 The fine structure of the world, the probability (or strength) of the interaction of quantum light with matter is determined by the fine-structure constant. Somehow, incredibly small values of fundamental constants are cancelled out and yield a simple and understandable number for human perception — 1/137 (although experimental data give a small correction to this value). Unlike many physical quantities, it has no dimension. It's simply a constant, like "π". Furthermore, if we let our imagination run wild, this constant could indicate some geometric description, or even may point to a human!

Imagine a photon for which space and time do not exist. Moreover, there is no cause and effect on the usual time arrow for this particle. And the world with its matter is just a strange "illusory substance", not how it is presented to humans. Try to imagine that you are a photon. Then, what exactly distinguishes you as a photon, that is, as a concept? Only that you exist. Maybe a photon is the elementary unit of information related to consciousness and the sense of "self-existence"? The information content of light has been recognized by some well-known scientists:

"It (light) is energy and it's also information-content, form and structure. ... It's the potential of everything. ... Light is this background which is all one but its information-content has the capacity for immense diversity. Light can carry information about the entire Universe", as David Bohn (1917 – 1992) explained (Weber 1990).

Many scientists and engineers in the 19th century admired the beauty of mathematics and the very concept of light quanta. For instance, Nikola Tesla (1856 – 1943) enthusiastically declared (Csanyi 2012) in 1899:

"Matter is created from the primal and eternal energy, which we know as light... Matter is the expression of infinite forms of light, for energy is older than it... Particles of light are notes... I am Light in human form... Just believe. Everything is light".

As you may remember from chapter 3.9 Idealism and information, the philosopher Gottfried Leibniz (1646 – 1716) believed that the foundation of our world is an infinite number of substances — monads. A monad is not a material or substantial entity. It is simple, indivisible, and has no extension (photon!). Each monad has a spiritual nature and is in constant change. And it is independent, that is, it exists by itself. Its changes occur spontaneously. Thanks to the continuity of existence, the monad is self-aware.

It must be acknowledged that this concept has been known since ancient times. The very concept of the monad was used by the Pythagoreans in the 6th century BCE, who named the first arisen being a "monad". For many Greek philosophers, including Pythagoras, Xenophanes, Plato, and Aristotle, the monad was a term for God or the singular source.

Photons with infinitely small wavelengths existed in the first seconds of the Big Bang. Their incredible density and quantity imply an incredibly large amount of information if there was some design for this creation. It was these high-energy photons losing energy (or cooling) that began to transform into elementary particles, then into hot gas, and subsequently, into stars.

When the Bible talks about uncreated Light (with a capital L) from the very first lines, could it be referring to this, the infinitely small wave limit of photons? This is "another light", not created by the Sun or the stars. The Big Bang was simply the gates, slightly opened from the world of ideas and forms to create matter. Our Universe, about 14 billion years ago, was in a state of cosmological singularity. It consisted of pure radiation with incredibly high energy (incredibly short wavelength of photons).

Our philosophical assumption resonates with many authors of the Agni Yoga movement, which was founded by thinkers Nicholas (1874 – 1947) and Helena (1879 – 1955) Roerich. Like many at that time, they believed in the world of "subtle matter". It is called like this because it is made of high-frequency vibrations. The higher the frequency of these vibrations, the closer we are to the subtle world, and the closer we approach the Light. The subtle world is a repository of information brought about by souls after the death of a person. In our reasoning, we can interpret the "frequency of vibrations" quite literally — it is the highest limit of photon frequency that existed at the time of the Big Bang. Remember, high-energy photons can give birth to matter, and thus our Universe.

But what is ordinary (created) light? Ordinary light is radiation visible to our eyes. It is the second type of light. It is much less energetic, with a lower frequency, and it appeared much later in the history of the Universe. This radiation is created by the Sun and other light-emitting sources. These low-frequency photons, bouncing from surfaces of various material things, are used by humans to receive information. The Bible talks about this light later, mentioning luminaries and the Sun.

We can go much further and ask — where is this subtle world, created from photons with incredibly high frequency and density? Perhaps, it exists in the past for us. This is the Big Bang. This subtle world (or singularity) appears to us in the distant past — about 14 billion years ago. We move along the arrow of time, and it seems that we are moving away from that singularity. For us, the Big Bang appears in the past because time is a concept exclusive to our world. For the existence and transformation of matter, time with its cause-and-effect relationships is mandatory. While for matter and us this moment of "creation" is in the past, time and space do not exist for photons themselves and for the singularity. The world of the Big Bang is outside our time — it exists in the past, present and future.

Perhaps the cosmological singularity is precisely the subtle world created from pure information. It is exactly this very beginning of the world that was characterized by the maximum amount of information, when the entropy of our Universe was at its lowest (see chapter 3.1 Entropy). This singularity consisted of incredibly energetic photons, each of which is an elementary unit of meaning. And their astronomical number gives birth to a complex reality of spirituality.

This is a world of pure spirit without matter and time. *"In him was life, and that life was the light of all mankind."* (John 1:1-18).

Although all this looks quite logical and consistent, I don't think the initial cosmological singularity can be described in the scientific concepts. This limit of incredible photon densities does not obey physical laws. However, like our narrative above, it cannot be confined within the boundaries of empirical sciences.

15 Knowledge through eclipses

Among the popular interpretations of extraordinary natural phenomena that cannot be explained from a scientific point of view is the short word — coincidence. If we are unable to answer a question using known laws, then the most naturalistically correct explanation is this: Anything can happen with a small probability, and we just happened to be "at the right place at the right time" to observe some coincidence. Since many events occur in different places around the globe at any given time, there must always be unusual and rare incidents. The human brain is gifted at finding coincidences in the millions of situations we deal with daily. Anything can happen, sometime and somewhere. After such an explanation, other reasons are usually not sought.

As we have already mentioned, there are events and circumstances in the world for which "coincidence" and "chance" are exceptionally absurd explanations. Such phenomena do not fall into the thousands of not very important events we deal with every day. If, for example, clouds covered the Sun across the entire territory of one continent, precisely for 6 minutes and exactly at noon, and then the Sun shone again, you would not accept "chance" as a reasonable explanation. Almost certainly, you would start looking for the real reason.

Coincidence. That is exactly the answer you will get when inquiring about why we can observe solar eclipses. If there is no good explanation, it is the most correct answer for science. But this book goes beyond science's boundaries.

15.1 Solar eclipses

A solar eclipse is a phenomenon during which the Moon completely or partially covers the solar disk for some time. The Moon, being between the Earth and the Sun, casts a shadow on the Earth on a specific area, from where this phenomenon can be observed.

The reason for such a phenomenon is in the incredible coincidence: The diameter of the Moon is about 400 times smaller than the diameter of the Sun. At the same time, the Moon is approximately 400 times closer to the Earth than the Sun. Therefore, from the Earth, the Moon and the Sun seem to be approximately the same size. Many think that there is no other explanation than sheer coincidence. Is that so?

One of the reasons why you might start asking such a question is because we know that there are about 290 moons and satellites of other planets in the solar system (Howells 2023). They do not create eclipses with the precision with which this happens on Earth. This means that the probability of such a coincidence related to a total solar eclipse by our Moon is (1/290) = 0.0034, at least if we assume that the distance to other planets and their sizes do not matter much for this estimate.

This feature leading to eclipses is compounded by another strange coincidence: The Moon is slowly moving away from the Earth (about 4 centimeters a year). Therefore, in the past, the Moon appeared much larger than the Sun, and eclipses were not as impressive. Obviously, the simplest invertebrate life forms can hardly be impressed by anything in the sky. For them, by and large, it makes no difference whether there is an eclipse or not. In the distant past, the Moon completely covered the solar disk. The "adjustment" of the Moon's size to make an eclipse possible occurred approximately 600 million years ago. This period is very close in time to the "Cambrian explosion", which was marked by the unprecedented appearance of complex organisms between 540 – 530 million years ago at the beginning of the Cambrian period. This event was accompanied by the emergence of many major types of living beings that make up the modern animal world. At that moment, many new major evolutionary branches of animals emerged. Such an event has never happened again, neither before nor after.

Why is it that precisely at the moment when complex animal life appeared, observing total solar eclipses became possible? After all, this is not an insignificant phenomenon among millions of others where chance rules?

As we have already mentioned, it is true that people notice all sorts of coincidences from millions of completely insignificant events and instances. But with the Earth, Moon and Sun, the situation is entirely different. These are the main significant objects for the existence of life and humans. They are privileged choices for our attention out of a vast number of insignificant things and phenomena. The sizes of the Earth, Moon and Sun, as well as the distances between them, have already undergone fine-tuning in such a way that life could exist. Any additional coincidences, such as the equality of the two instances of the number 400 for the occurrence of eclipses, precisely in the period when conditions for the emergence of an intelligent observer began to appear, are statistically improbable. Here, we observe a dual coincidence in both spatial (distances and sizes) and temporal (time-based) aspects.

As we explained earlier, when dealing with unlikely coincidences without causal connections, we need a "filter" to narrow down the vast number of people, events and many other things to a small statistical sample. It is precisely through this way that one can notice interesting phenomena that are difficult to interpret when dealing with a vast number of possibilities. This is what we did when looking for coincidences in significant events in the lives of famous people (see chapter 5 How to crack the code).

The fact that the Earth, Moon and Sun are the most significant categories for life and humans means that any coincidence with such categories is unlikely to be random. The only question is whether there is some design in this apparent coincidence. Those who believe in the intelligent design of our solar system often follow this line of thinking: The reason for the precise adjustment of the Moon's size and its distance from the Sun is to help humans understand the world.

Indeed, in the past, solar eclipses were used for the most important scientific discoveries. For example, solar eclipses are essential for observation of the solar corona, which allowed the discovery of helium. This incredibly important monoatomic gas is colorless and odorless. In terms of importance, it ranks second after hydrogen. As we now know, hydrogen and helium make up the bulk of the mass of many stars, including our Sun. Helium was discovered by French astronomer Pierre Janssen (1842 – 1907) during the solar eclipse of 1868 in India. The total phase of the eclipse lasted about 6 minutes. The astronomer discovered a bright yellow spectral line in the solar prominences at a wavelength of 587 nm (see chapter 14.1 Little bit of science). Subsequently, this element was named "helium". It was so named because "Helios" in Greek means "Sun". It is helium that creates radiation at this wavelength of the electromagnetic wave.

Thanks to the 1919 eclipse, the General Theory of Relativity, arguably one of humanity's most significant discoveries, was experimentally confirmed. It was theoretically proposed by Einstein back in 1915. In his paper, he asserted that gravity is not a force of attraction acting between bodies in space, as Isaac Newton had previously explained, but a property of space-time. This theory states that gravity arises when the surrounding space-time is curved under the mass of any body. To test this hypothesis, English astronomer Arthur Eddington (1882 – 1944) organized expeditions to observe the solar eclipse of 1919 in Brazil. The total phase of the eclipse lasted about 6 minutes (precisely, 6 minutes and 51 seconds). This eclipse was the longest in the preceding 500 years. The observation results convincingly confirmed Einstein's General Theory of Relativity prediction about the deflection of light in the gravitational field of the Sun. Gravity can indeed deflect even light!

If there is any higher reason behind a solar eclipse, it could be this: to give people the opportunity to accelerate their scientific and cognitive progress in finding new knowledge about this world and make discoveries. Such precise adjustment of the sizes of the Earth, Moon, Sun and the distance from the Earth to the Sun — all this was necessary to create the necessary conditions for biological life, allowing water to exist in a free state. But that was not enough. The distance from the Moon to the Earth was adjusted at the most necessary time, when complex intelligent beings would appear, so that they could make the most significant discoveries.

As you may have noticed, the maximum duration of the total phase of these two eclipses, during which both of the most important discoveries for humanity were made, was about 6 minutes. This is the result of an incredibly precise adjustment of the Moon's size, its distance from the Earth and its rotation speed. From previous chapters, we already know the significance of the number 6. This figure "penetrates" everything where there is a hint pointing to humanity.

Moreover, an eclipse is a hint at how to discover the very method that will allow the detection of the design of the Universe. And the method itself is quite simple. We used it in the previous chapters. When we are dealing with a massive stream of information that does not allow us to find some pattern or regularity from which one can learn something useful, the simplest thing is to focus on the important part of this information and "block" the main bulk of useless data (or background). The latter makes the subject of study unimaginably complex. This is exactly what we did throughout the book to reveal statistically improbable coincidences or synchronicities that cannot fit into the naturalistic explanation of the world.

A solar eclipse is precisely a demonstration of the principle mentioned above. The Moon, obscuring most of the solar disk, acts as a filter. It blocks billions of photons flying towards the Earth, which do not allow the Sun to be studied. However, we can easily investigate the solar plasma prominences after the Moon blocks the huge stream of information-carrying photons that make such studies impossible.

Thus, an eclipse, metaphorically speaking, is the key for discovering the very method of detecting synchronicity in chapter 5 How to crack the code. A key that was presented to us at the moment when life, including intelligent life in the form of humans, began to become self-aware and capable of analyzing the world. And this very key was created using the same principle — an unusual match in numbers, or the coincidence between the size of the Moon and the distance to the Sun, with a logical connection to a significant informational phenomenon. Namely, it happened at the moment of an explosive emergence of new information and the most powerful surge in evolution of biological organisms in the entire history of the Earth approximately 540 million years ago. This event was necessary for the creation of new complex types of flora and fauna. And subsequently, for the meaningful perception of the world by humans.

I promised that this book seeks meaning in observations that go beyond scientific explanations. The meaning of eclipses could be this: It is a hint to us that our world is created. Not just for life, but for intelligent life seeking answers. In the case of the coincidence of the two instances of 400 for the occurrence of eclipses, it is an exceptionally clever puzzle. It was created on a vast scale of space and time. As we know, when so much meaning is embedded in a coincidence, that coincidence ceases to be a random feature caused by pure chance.

15.2 Higgs boson

Let's consider another example of this same principle — blocking unnecessary information to discover new laws of nature. This time, we turn to the realm of the smallest scales ever studied by humans. We have already talked about this in chapter 7.2 The Law-giver.

The Higgs boson is a fundamental particle of the Higgs field, responsible for the existence of mass of the most elementary particles, which, at this point, cannot be broken down into smaller components. This field was first proposed in the mid-sixties by Peter Higgs, after whom the particle is named. As we have said before, this particle was finally discovered on July 4, 2012, by researchers at the Large Hadron Collider (LHC) — the world's most powerful particle accelerator, located at the European Organization for Nuclear Research (CERN) in Switzerland.

The LHC confirmed the existence of the Higgs field and the mechanism by which mass arises, thus completing the Standard model of particle physics — the best description of the subatomic world we currently have. Without the Higgs field, no elementary particle would have mass, meaning fundamental particles would fly through the Universe at the speed of light. This implies that there would be no stars, no planets, and no us. Due to the incredible importance of this particle, popular media has dubbed the Higgs boson the "God Particle". However, there is a particle that the Higgs field does not endow with mass. As you might have guessed, it is again the particle of light — the photon.

At the LHC, the Higgs boson was discovered by accelerating two protons to speeds close to that of light and then colliding with each other. This creates a cascade of other particles, which quickly decay into lighter particles. The Higgs boson decays too quickly to be detected directly. Among the various Higgs boson decays, there is one decay that is the most well-known and easily detectable — the decay of the Higgs boson into two photons. It was this decay that was used to discover the Higgs boson. To this day, it remains the most important and easily detectable decay, which scientists use to study the properties of this amazing particle.

To detect the decay of the Higgs boson into two photons, it was necessary to ignore billions of background events where photons are flying from the decay of other known particles. The photons emanating from the decay of Higgs bosons have very high energy — around 50 gigaelectronvolts (GeV) or 1 billion electronvolts. If you convert this energy into the length of an electromagnetic wave, it corresponds to about 2.48×10^{-8} nm (nanometer). These are incredibly short waves, as one nanometer equals one billionth of a meter. Visible radiation are electromagnetic waves with a wavelength of 380 – 780 nm, perceived by the human eye as light (see chapter 14.1 Little bit of science).

It should be noted that the Higgs boson does not interact with photons directly. Initially, a pair of other heavy particles is created, which then emits two photons. And of course, the fine-structure constant and the number "π" play a crucial role here. See chapter 9.7 The fine structure of the world. It turns out that the amplitude of such interactions is proportional to the square of the fine-structure constant and inversely proportional to "π" cubed. This pair of fundamental parameters appears every time a photon interacts with matter.

I hope this discovery of the Higgs boson using photons reminds you of something. As in the case of a solar eclipse, the same principle was applied here — blocking unnecessary information carried by photons, to help reveal truly important data to make discoveries. As we mentioned before, the same principle was used to detect meaningful coincidences in the lives of people after we ignored a vast amount of unnecessary data and isolated only the happenings that can be tracked and processed (chapter 5 How to crack the code).

15.3 Consciousness outside the body

The enormous flow of information our brain receives throughout the day prevents us from focusing on what may be absolutely central to our being. Our brain does not only receive information through the five senses: taste, smell, touch, hearing and sight. It also receives information from various organs of our body for physical existence among molecules. This second stream of information is not sought out by us. It fills our brain which, in turn, processes this data unconsciously. Based on these calculations, our consciousness makes the most optimal decisions for our behavior.

What if we block all these data streams from this world? Maybe then we will be open to the true reality beyond what we see around us? Maybe the realization of the true "I" arises precisely when the body and brain cease to function or reduce their activity? When our intellect stops functioning to process data from our biological clothing, leaving our consciousness alone? At such a moment, all that remains of us is our projection in the world of meaning and abstract forms. This is the most important part of our consciousness that makes decisions and launches intelligence algorithms to control the biological shell made of molecules on Earth. By blocking the information streams from our body's signaling system, we may get the opportunity to come into contact with the world of Light, the true homeland of our soul (chapter 14.3 Little bit of fantasy). This is an interesting idea for contemplation.

You probably could hear a lot about such amazing experiences occurring in various situations when the body is in an extreme state, between life and death, and the sensory capabilities of the body and brain are blocked, intentionally (or unintentionally) altered or somehow muted. Sometimes this leads to the experience of leaving the body. An out-of-body experience is when you realize yourself to be outside of your biological shell. At the same time, you can observe your inactive body from some distance. According to people who have had such an experience, it is like the soul leaves the body and freely wanders in space. Psychoactive substances often lead to out-of-body experiences and are frequently used in the ritual practices of various cultures. In addition, some people can also leave their bodies consciously (Щербаков 2021). For example, the well-known theoretical physicist Richard Feynman (1918 – 1988) experienced out-of-body experiences at will during experiments in a saltwater tank isolated from external influences (Feynman and Leighton 1997).

The sensation of leaving the body is a characteristic element of near-death experiences, which occur in life-threatening situations or clinical death. One of the earliest descriptions of this phenomenon is documented by Plato in "The Republic", published at around 375 BC. It is Plato's teaching that justifies the body as merely a temporary shelter for the immortal soul. It should be noted that such an attitude towards life and death was common for almost all ancient cultures.

Since then, hundreds of descriptions of the out-of-body phenomenon have been published. According to various estimates, these experiences occur in at least 5-10% of people. This phenomenon is increasingly receiving attention in scientific research (Sellers 2017). In my opinion, it is completely foolish to deny the phenomenon of out-of-body experience. I won't even try to convince the reader that this phenomenon is a reality we are dealing with.

Many people who have experienced clinical death talk about similar visions. They are able to see their body when doctors are trying to restart their heart. Then, a dark tunnel appears, at the end of which a bright light illuminates. Most of those who go through this tunnel find themselves in another world, bathed in unimaginable bright light.

Some describe meetings with deceased relatives or a disembodied be-ing radiating kindness. They say that they experienced a sense of un-precedented peace and incredible love. A moment of euphoria comes, associated with the feeling that they have returned to their real home, from which they came from to Earth. A feeling of incredible knowledge and pure awareness appears. They begin to see colors for which there are no descriptions. They start to have a 360-degree vi-sion. Space does not exist in such moments. And time has stopped, or it ceases to be felt. Then they are made to understand that their time to leave Earth has not yet come, and they need to return. Communication occurs mentally. And upon returning to life, such people fundamen-tally change, becoming more spiritual.

From a medical standpoint, there are several naturalistic ex-planations. These relate to the brain's oxygen starvation or purely chemical processes occurring in the brain under the influence of med-ications. Essentially, all scientific explanations boil down to one thing: During the pre-death period, different areas of the brain continue to function in coordination, even when no oxygen is supplied to the brain. If the heart stops, then blood no longer reaches the brain. But some processes in the brain still occur even in the absence of heartbeat and circulation. It is precisely the chaotic surges of neuron activity during clinical death that can cause such paranormal visually-vivid stories.

Of course, as is often the case with science, models dealing with mechanical processes cannot provide a full explanation of the subjective experience that people undergo. From an informational per-spective, such bursts of neuronal activity in the dying brain should lead to a very wide range of visions. They could be anything. They could be randomly flashing images, bursts of memories, or some complete nonsense. For example, a flashing number 66. But this is not what people experience. The overwhelming majority of accounts of what was seen during clinical death are incredibly similar. And they make sense too.

I will not delve into the topic of clinical death. This chapter is simply an invitation to review the relevant literature. All I can say is that the near-death experience is an intriguing confirmation of the material from the previous chapters. Specifically, in such critical moments, being on the brink of death, people gain access to new information. This moment of access coincides with the blocking of the vast flow of information received by our senses tuned to this material world. It is similar to an eclipse. The meaning of this new information goes far beyond what we deal with in this Universe. All attributes of the world of ideas and abstract forms are clearly present in the visions of people during such near-death moments: the absence of time and space, incredible experiences, euphoria from being in one's true home, meeting with the departed, and bright light of unknown origin. Why light? We have discussed this in chapter 14 Light.

16 Beyond this reality

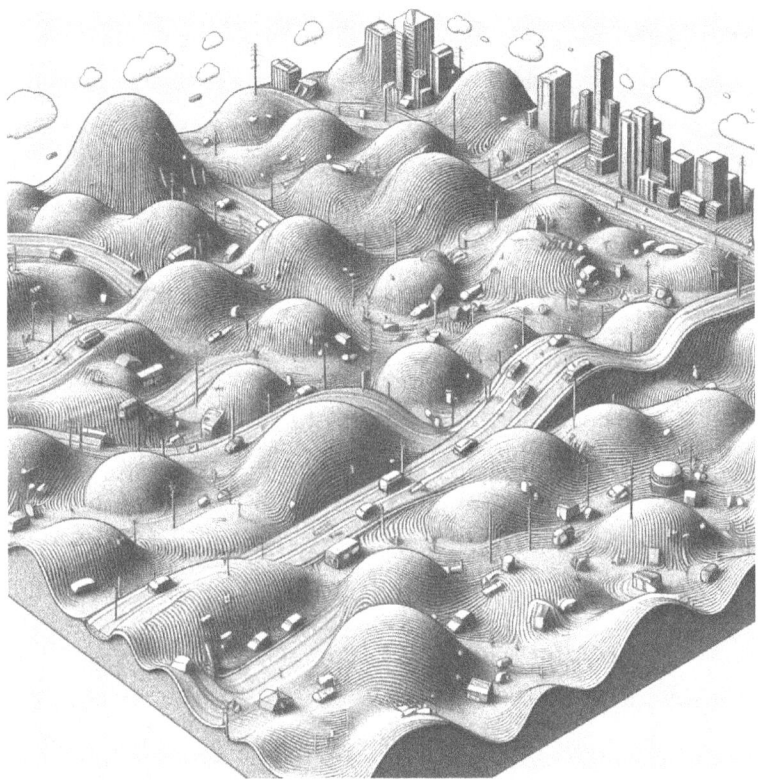

Unusual happenings in the lives of people analyzed in chapter <u>6 Examples of significant coincidences</u> are examples of significant coincidences that are difficult to explain as mere chance. We admit that one or two strange coincidences are possible. But finding 6 instances of synchronicity among people whose names are well-known and familiar to many from school is quite unusual. We even obtained an estimate of the probability of discovering a previously known group of 6 people with whom synchronicity occurred due to blind chance. This probability, found in chapter <u>6.9 Nothing is accidental</u> turned out to be astronomically small — of the order of 10^{-11}. Such phenomena may reveal a profound unity of mind and matter, as well as subjective and objective reality, as Jung and Pauli were convinced.

It's likely that some of you have had the luck of picking cranberries in a swamp. When I was little, my father and I often went to peat lakes 30 kilometers from Minsk. These were fields overgrown with a thick layer of grass, moss, small bushes and birches. The red cranberries grew in solitary clearings among the bushes. When the berries ripened, they could be found lying in the soft moss. Once, during one of such trips, I heard a crack and looked around. Somewhere in the distance, almost simultaneously, two rotten birches fell. There was a considerable distance between them. The first question that came to my mind: How is this possible? Two rare events occurring at the same moment? I only realized what actually happened a few seconds later, after feeling the vibrations of the moss under my feet. It turned out that someone had walked close to these rotten birches. This caused the grassy cover to move in oscillation, leading to the falling of the rotten trees.

The earthen surface in dry lakes, where peat begins to form, is not solid. It is a mat of intertwined roots, beneath which lies the aquatic world of a dying lake. It is completely different from our world. It has its own ecosystem and life forms. The world on the surface is entirely dependent on these layers of water. When someone walks or jumps on this carpet of grass, everything around moves in a wave-like motion. This causes events that can be interpreted as incredible coincidences. And even the world of water itself, by emitting gas bubbles, can affect everything on the grassy surface.

Imagine what would happen if you called someone. Perhaps if you did this, you would see trees falling simultaneously or bushes rustling. These events might seem like coincidences to you. Then, you would feel vibrations under your feet, and only after that would you see the person coming towards you.

Can such an analogy work to explain synchronicity? If so, then the world that animates the atoms and molecules of our reality does not have an arrow of time. Or this arrow is pointed in a completely different direction. Therefore, the influence of that world does not obey the cause-and-effect relationships in our Universe. These waves can spread both in space and in time. Moreover, they affect the destinies of people, revealing that mysterious reality to us.

What does this mean and how can it be illustrated? In chapters 3.9 Idealism and information and 15.2 Higgs boson, I have already provided an analogy related to the Higgs field. To find something important, one must "excite" matter. For this, two particles are accelerated nearly to the speed of light and then collided. Receiving highly concentrated energy, matter is excited, creating ripples in its wave structure. Many new particles appear. Then, to detect what interests you (the Higgs boson), you need to block the main flow of information (the photon stream) that is irrelevant to your goal of enquiry.

If we use such an analogy, perhaps the unreasonable coincidences at the most critical moments in people's lives are the result of some kind of "wave ripples" in the space of events and associated information. Human emotions are a source of excitement for semantic information, which lies at the foundation of the world of Platonic abstract forms and ideas. When this happens, coincidences occur, that is, waves in events, in the system of concepts. This wave ripple compresses and expands the "space" of events. Ultimately, this leads to repetitions and coincidences in people's lives. If the model of semantic structure was built based on logic and mathematics, then it influences the formation of coincidences in various numbers, dates of birth and death, and any other numerical values that relate to people.

To detect such waves in the structure of meanings, events and their characteristics, it is necessary to block information from a vast number of people and exclude coincidences that are not very significant. It is necessary to focus on a small group of predefined people and their most significant events. See chapter 5 How to crack the code.

Here is one of the possible analogies: Imagine a stretched rubber film. Or an elastic membrane. People (or human consciousness) are nodes or bulges formed on the surface of this film, which slightly protrude upwards. The illustration at the beginning of this chapter provides a vivid illustration. Such bulges are made from the film itself. The film and everything beneath it are the semantic information of the Universe — the Platonic world of abstract meanings and ideas. The film is sprinkled with mountains, rivers, houses — this is the material world. All these material objects are not attached but are just resting on the film. These projections of ideas, created from inanimate matter, can move, roll over, change their location in space (on the film) and in time. The nodes representing human consciousness can interact with all these material objects. All these physical objects obey physical laws, which determine how objects can move on the film, collide, and sometimes interact with the nodes-people. These natural laws are fully defined by this film, but we can only see their shadows and can deduce small fragments of such laws by observing material objects and processes. This is a picture of the world where matter, although supported by the film, is secondary.

Now imagine that a person experiences a pivotal event in their life. This is precisely the genuine effect of their consciousness and emotion, rather than intelligence. The film begins to oscillate around this node-person, leading to circular waves on the film. These oscillations are a surge of semantically important concepts connected with our emotions. As a result, other nodes begin to oscillate. For example, this affects people who are close (again, we mean close in terms of meaning and semantic relation). As the film oscillates, material objects around such a node come into motion too. Being merely projections of oscillating abstract forms, they are not attached to the rubber film. They have no consciousness. They can have their oscillations in space and time, not always coinciding with the behavior of abstract forms. One must not forget that matter has its degrees of freedom and can interact with itself. Pure chance also plays a role.

This example is simply an attempt to uncover the interaction between mental and physical events through a semantic model of being. Improbable coincidences are oscillations caused by circumstances which are significant to a person's psyche. The emergence of amazing coincidences in numbers, names, birth dates and death dates also occur because the film oscillates like circular waves on the surface of water, after we have thrown a stone. Such waves lead to repetitions in semantically related characteristics of events.

Since the film is created from concepts representing meanings, then space and time become secondary. Temporal coincidences, such as coincidences in dates of events which are close in meaning, are possible, just as spatial coincidences are. In the universal world of meanings, time, as such, does not exist. Nor does space. These are just concepts necessary for the operation of physical laws and for the interaction of matter with itself and with people.

We live in a moment of time when the unusually rapid progress in science and technology begins to "gnaw through" the spatial-temporal covering stretched over the framework of abstract meanings. On this framework rest the decorations of the material world with which we deal every day. Quantum mechanics is one example where the logic and intuition arising in the world of objects, with which our material shell directly interacts with, begin to falter.

351

These coincidental events affect not only the fates of people, their significant events and their timing. They change the material world and the structure and events of the past, which becomes "elastic". Past events can "deform" depending on the present.

In the physics of the microworld, the concept of nonlocality and entanglement of particles is firmly established. The concept of "retrocausality" in particle behavior suggests that there is a mechanism that allows present circumstances to correlate with their past states, see chapter <u>8.2 Nonlocality</u>. At present, retrocausality is a hypothesis for explaining quantum mechanical phenomena. It suggests that the state of a quantum system in the present can "edit" information about the past. This sounds extraordinary. Nevertheless, this hypothesis is no worse than the hypothesis of an infinite number of universes in which the observers continuously "flows" from one universe to another, depending on what they observe.

We do not know what the past was like. We cannot observe, measure or experiment with it. But there is information that has reached the observer. It is this information that synchronizes with the present. Of course, such phenomena are considered for microscopic particles. For larger objects (and a human being is a large "object"), the probability of such phenomena is negligible. But for physical objects — a small probability multiplied by a huge number of their parts and everyday situations could potentially make very rare events possible. What if the brain has a very direct connection with quantum reality (Penrose 1989) which increases the retrocausality behavior, especially in moments of strong emotional experiences of consciousness?

Real-life examples related to the adjustments and correlations in people's extraordinary circumstances have already been shown in this book. See chapters <u>6.6 The Kaiser and the War</u> or <u>6.7 Again about</u>

wars, where we tried to explain the discovered coincidences with the effects of editing the information of the past. Even remarkable coincidences found in ancient superstructures might be explained by the peculiarity of our consciousness participating in retrocausality, see chapter 9.9 The Great Pyramid and numbers.

Here is another fantastic example: As I've already mentioned, the Higgs boson was discovered in 2012 at CERN. For many scientists, this meant the culmination of their careers as physicists. Billions of dollars were spent on the experiment, and the search went on for decades. What if the information about the past was correlated with the present in such a way that Higgs' paper was accepted for publication in 1964? Higgs' original paper, where he predicted a new particle, was not accepted for publication initially because it was not clear how this paper related to reality. However, he decided to send it to another journal, almost without changes, and the paper was accepted. What if this was an altered piece of information from the past to enhance the emotional state from the events of 2012 — the discovery of the Higgs boson, which everyone so eagerly wanted to find? Someone or something decided – "Alright, here's your Higgs particle, with the mass you want. You've got it. Just let me adjust the past in 1964 and arrange for this predictive article to be published. Otherwise, how will you experience the great emotional pleasure of its discovery?"

Of course, we have no proof whatsoever that these events happened exactly like this. To justify such a sequence of events, it is necessary to find some coincidences in meaning, as we did in 6 Examples of significant coincidences, and ensure a low probability for such astonishing incidents. The discovery of the Higgs boson is not so significant for human lives, compared to wars and life and death situations, that synchronicity could emerge. But if it did, the real story may look as we have just described it.

Speaking of retrocausality, one might ask the following question: Suppose information about the past is edited to satisfy the strong emotional experiences of people in the present. But how is this possible? Do all records in the past change simultaneously, all mentions in books, all inscriptions on monuments, and even signs on buildings? If what changes all of this exists outside the realm of time, then there is no contradiction. Those who program know this simple fact: To change something in a computer game, it's enough to change one software object. All other objects associated with it change synchronously, both in the past and in the present of that game. Thus, we are dealing with one universe, not with an infinite number of them, as is the case in some interpretations of quantum mechanics. In our universe, the past is being adjusted to the present. We need just one universe, not many. This is a much more efficient way for the "mechanism" of our reality than creating an infinite number of universes existing in parallel. As we have explained, the many-worlds concept of quantum mechanics is too complex to be true and fundamentally untestable.

You might ask yourself: Okay, we've established all these improbable coincidences in chapter 6 Examples of significant coincidences. It is unlikely that these effects are due to randomness. There are no causal relationships between them. But what do they mean? They appear to us as spontaneous anomalies in the lives of individual people, groups of people and even entire countries. These events manifest in various numbers related to people's lives, spanning large distances and long periods of time. Such unusual phenomena adjust physical parameters on the scale of the entire Universe (chapter 2 This world is such because we are in it) and, possibly, increase the chances for the emergence of life from inanimate matter (chapter 3.5 Life and information). It would not be surprising if they can accelerate evolutionary processes for the emergence of complex organisms from the simplest cells (chapter 3.6 Theory of evolution). One can only specu-late why such phenomena occur. Here are two possible explanations:

1. This is merely the glimmers of a conscious yet indifferent world of abstract forms standing behind material reality. It's simply un-desirable irregularities in the fabric of information used in "paint-ing" the picture of our Universe. Perhaps these are some kind of errors in the simulation, or some anomalous fluctuations in the transcendental information of some unconscious mind (if an un-conscious mind is possible!), which is indifferent and cares neither for humanity nor for individuals. Maybe it's just a by-product of the unconscious "will to live", as described by one of the most famous German thinkers — Arthur Schopenhauer (1788 – 1860). According to his philosophy, will is a blind, irrational principle in the otherworldly essence of the world. We will consider this hy-pothesis later.

2. Or they are mysteries or well-thought-out "clues" given to us from another plane of (immaterial) reality. Something or someone is trying to attract our attention. It offers keys that unlock the gate to something entirely new. It invites us, without coercion. But this invitation is only for those who understand where to find these gates. Perhaps we can use them to change our lives. This is an informational reality that cares about us. It is also a loving mind that changes us for the better. Once the key to this reality is found, we will also be able to find the gate and advance our entire civili-zation forward. Or, perhaps, these clues come from ourselves, from that hidden reality, since sooner or later we will all find our-selves in it.

I am inclined to say that the second explanation is the most probable. In several examples in this book, I have shown that an invisible informational reality is not indifferent to our destinies. There are many reasons to think that people have begun to decipher this reality, starting with the widespread acceptance of religion.

16.1 God of meaning and truth

Following everything said above, we can assume that God is the informational reality of truth and meaning. This is the highest form of information. In our world, the concept of information requires a rational creator and recipient of messages. For the material world, information is not something natural. That is why it's necessary to transform messages carrying some meaning into data, in order to preserve them later on material carriers, such as books and computer disks.

Unlike us, God is both the creator and recipient of information. This implies his transcendence (from Latin "transcendens" – going beyond limits) – reflecting a realm beyond this world. I remind again that this book uses "he" as a pronoun when referring to God, for simplicity. The large oval in Figure 3 in chapter 3 Information, meaning and consciousness illustrates this concept. The reality of God is truth and meaning, which also includes beauty and love. These two concepts have briefly been discussed in this book; however, many can infer such notions by relying on their inner sensations. There are numerous testimonies of people who have experienced out-of-body experiences during clinical death, who can very accurately describe (Moody 2005) that beauty, light and love are unconditional attributes of a reality beyond our world. I talked about this in chapter 15.3 Consciousness outside the body. Many philosophers, from Plato to modern authors, believe that beauty is an integral part of God. Not coincidentally, some thinkers use the principle of beauty to prove his existence (Swinburne 2004).

The concept of God as pure information about meaning and truth is not something new. Interestingly, from the perspective of the information approach, this concept of God aligns well with traditional philosophical and religious teachings.

How can one imagine God as a world of abstract ideas, existing without material carriers of information, where there are no time, space and matter? I don't think humans have sufficient intuition to imagine that reality. We do not deal with immaterial carriers of information in our everyday lives. But perhaps some analogies can help. Imagine an endless ocean (although even here, it's hard to imagine infinity). Each "molecule" of such an ocean is a concept, an abstract model of something. These concept-molecules do not occupy space and exist outside (our) time. If you wanted to find the home of your childhood – it is already there. It exists because the concept has been "detailed" informationally by all the people who once lived in it in our world, such as your parents, relatives, or neighbors, dead or alive. It is also created by yourselves from memory. You share this information during sleep. Possibly, in that reality, you can find everything that people and animals experienced while on Earth.

Perhaps everything from our Universe, which requires information and animation, is already in that hidden world. As Plato famously stated, we are but shadows. It is often said that if there is a God, why did he create an infinitely vast amount of matter, stars, galaxies and flying rocks in incomprehensibly huge but almost empty space? It does not seem very rational if the goal was to create life here, on Earth. Maybe this is the answer: Creating a universe takes as many resources as the resources needed for one snail. The fact is that the concepts forming our material Universe do not occupy any space and do not require energy in the world of abstract meanings. God did not need to create every tiny detail, as a few basic rules and laws were enough for all the details of our Universe to emerge on their own. Having an algorithm is sufficient to create an infinitely vast environment of things. We illustrated this in chapter 9.8 The infinite beauty of fractals.

The world of meaning and the substance of ideas is primary. It has always existed. It was created by no one. Its foundation is goodness and love, simply because evil cannot be the creating principle of anything. For many, the fact that we see this world around us indicates that creation is the foundation of that reality, thus it is a manifestation of goodness. Evil is a concept for destruction. The mere existence of something already speaks of a creative beginning of our reality.

Everything stated above is simply conjecture, based on the principle that there must be a first cause for our world. This is what the ancient Greeks called the foundation of being. And they added that thought itself is identical to being.

16.2 Knowledge

The reader might ask this question: Are knowledge and God intertwined? Often, it is said that God is all-knowing. According to religious teachings, he possesses all existing information, knows all secrets, and has the answers to all questions. But then, what is the purpose of living if everything is already known to God?

I think that if humans could glimpse into the reality of meanings, they would hardly perceive it as mere knowledge. In our understanding, knowledge is a description of how everything is arranged in this material Universe. But such knowledge is much more superficial than compared to the level where our world was designed, if that happened at all. There, completely different algorithms and laws operate.

Perhaps we can only understand some mathematical constructions which were used to develop physical laws (Penrose 2007). We can only observe "shadows and traces" of such abstract structures in mathematics and some laws of nature, which are striking in their beauty and logic (see chapter 9 Coincidences in numbers).

Here, one can again draw an analogy with a computer simulation. A computer character might gain knowledge about how their reality is structured, however, the true knowledge of how all this works is beyond their access. The world of this computer character was conceived and created by algorithms that have nothing in common with the knowledge gained by this personality by observing their surroundings. Even if it would be possible to deduce the law of gravity by observing falling objects inside this game, this person would never figure out all the details of implementation of gravitational phenomena in the software algorithm and in the hardware of this computer. Any circumstances that has led to the design of this game will be utterly inaccessible for comprehension.

If everything up to this point seems plausible, then why don't people who have experienced near-death moments bring back useful information from the reality they visited? We discussed this in chapter 15.3 Consciousness outside the body. For example, one could learn about all the laws of nature while in a state of clinical death. There are many testimonies out there that in such a state, people begin to very clearly realize all the knowledge inaccessible to us on Earth. Why not share it with the world upon returning to full consciousness? To my knowledge, this has not happened. In the overwhelming majority of cases, dreams also do not provide us with any useful knowledge, except for rare exceptions.

The reason might be this: We already said that our world is likely created using completely different algorithms than those we can study by observing the world in our dreams or in clinical death. Even if we receive them in a dream, we will not be able to interpret them using the categories of our world. A programming code written in a certain programming language cannot easily be understood by a character in a computer game whose goals of existence are entirely different. The beautiful mathematical formulations and laws that a person discovers while engaging in science are just some components of an incredibly complex creation. We discover them as separate parts of an intricate machine. They are inconceivably logical, beautiful and polished. But they are just unrelated parts. They may possess some functionality, but we lack a comprehensive understanding of how all these components integrate into the grand design of the Universe.

Secondly, how do you imagine a world where knowledge can be acquired during sleep or clinical death? Where there's no need to learn? And where the existence of God will be so obvious that it will take away a huge part of our free will, as we'll discuss later in chapter 17 Free will. The absence of direct access to such information is a perfectly logical consequence if our world was deliberately planned and created. There simply aren't any other options.

Now we can summarize everything here. Our knowledge reflects the laws of nature, which do not depend on us. It can be assumed that they are the rough "outlines" of an incredibly complex design involving matter, energy, space and time. However, these laws were created using more fundamental abstract concepts. Mathematics and logic are just the "shadows" of those concepts, which are still inaccessible to us. But science can move us closer to their understanding.

16.3 How to understand synchronicity

The phenomenon of synchronicity, discussed in chapter <u>6 Examples of significant coincidences,</u> suggests that this world does not always obey cause and effect relationships, which are the backbone of the natural processes. Putting this in other words, all these rare coincidences observed in people's fates do not have a direct causal nature. Their occurrence due to random chance is also statistically improbable. Thus, they require a different explanation.

Such phenomena would not be very surprising if they occurred on small scales described by quantum mechanics. However, synchronicity phenomena are observed on larger scales, and are often connected with people and their experiences.

In my view, these phenomena indicate that our world is within some dynamic timeless system that has an influence on it. And the dynamics of this system are not determined solely by the laws of the environment in which we find ourselves. Our material world, including humans, can be envisioned as continents on Earth's surface, floating on the mantle, and are always in motion. They are capable of colliding, encountering each other, oscillating and deforming, like tectonic plates. This leads to synchronicity events. Understanding these events is not easy, as the mechanism describing their movement and dynamics does not follow the usual natural laws. They are beyond our understanding. Perhaps we are in the presence of God?

I don't know why, but the analogy with a computer simulation often proves remarkably convenient in our discussions. Being a character in a computer game, you don't have access to the code of the game. Nevertheless, you can alter the gameplay itself. In cases of strong emotional stress and important events, information flows undergo greater optimization. For instance, when events in the game unfold very rapidly, the computer processor heats up. The cooling system can't keep up, and it overheats. To prevent this, it's important to optimize the execution of the program to enhance its performance. Special algorithms are activated. They are carefully designed "shortcuts" within the program logic, which remove unnecessary details and simplify the execution of tasks related to creation of scene-events.

For example, the episodes described in chapters 6.1 Abraham Lincoln and John Kennedy and 6.5 Hitler and Napoleon seem as if the events were optimized in such a way that the fates of people who had a significant impact on history appear quite similar. Perhaps the idea was this: Since the people involved in these events have a lot in common, similar elements could be used in creating their destinies. This is much more sensible than creating new situations in their lives and in history each time. Chapter 6.4 Gravediggers of the USSR, describing the end of the path of people who influenced the destruction of a huge country, may be a consequence of such an optimization. What was the ultimate goal? Perhaps the main purpose was to quickly end the preceding era and definitively turn the page of history in order to move forward.

Of course, it is difficult to say that we can thoroughly understand all these unusual phenomena. The world of meaningful information is not accessible to our intellect, as it is optimized only for interaction with our Universe.

16.4 Symbols and forms of ideas

If our material world is a construction of atoms, based on meaningful input information, then what is the language of this information? Could it be bits — zeros and ones, like in a regular computer?

Or perhaps qubits, which are in superposition of these states, meaning they can simultaneously take values of 0 and 1? Qubits utilize quantum mechanics phenomena to process data. In the future, qubits may enable the creation of ultrafast quantum computers.

We do not know what conveys meanings. However, there are intriguing guesses. It is possible that the language, on which the entire "mechanism" animating our world of atoms is built upon, might be most naturally interpreted by us in the form of symbols. They are the most compact way to express meaningful information. Of course, this does not mean that symbols and signs are the actual language of that hidden reality. Symbols and signs, as we interpret them, belong to our world. However, those abstract forms on which our Universe is constructed might easily "project" into our world and be perceived by us as signs. This is because they are the most natural representation of the ultimate meaning.

Assume our guess is right. The true language of meanings, symbols, are likely similar to the forms of abstract ideas. Here, we come back to the teachings of the Athenian philosopher Plato. He proposed there is a corresponding abstract form for any object of this Universe. Plato's world of ideas spurred on idealism, as we discussed in chapter 3.9 Idealism and information. I don't think that, at this moment, we have a good understanding of how idea-forms can exist independently, that is, without biological carriers.

Historically, the emergence of writing was associated with the activity of deities. Signs and symbols were the most accessible to our cognition. They were the most natural way to express thoughts and feelings. This is why the first writing appeared as pictographic symbols and hieroglyphs. Symbols always had a sacred meaning for primitive people. For them, for instance, the symbol of a circle meant the

Sun. What can be simpler and more natural? As we know, symbols were heavily used in all sacred texts.

Let's assume that abstract forms are projected into this world via our minds as symbols. But how do symbols relate to meaning, that is, to the abstract forms? This is an extensive area of contemporary research.

Many researchers have dedicated a lot of work to symbols. For Carl Jung, symbols represent something that carries the deepest mystery. A symbol *"enters human consciousness organically and becomes something familiar and personal"* (Nalimov and Drogalina 1995). Perhaps this is the natural language of our consciousness, which builds projections of the world in our heads? I'll try to hypothesize the whole idea like this: Information about sounds, objects, smells and tastes reaches the brain. Our intellect processes the information from the surrounding world and directs it to the consciousness in the form of symbols. Conversely, symbols processed by consciousness return through the intellect in the form of our actions. Our consciousness likely operates with the help of the language of symbols. They are a high-level code for creating our dreams in the subconscious.

Austrian psychoanalyst Sigmund Freud (1856 – 1939), the founder of psychoanalysis, believed that our dreams use symbolism to form dreamscapes. During sleep, our thoughts transform into abstract symbols. This transformation of data includes compression and extraction of the most important parts of our experience.

However, Carl Jung disagreed with this. He believed that symbols are primary and are the main product of the unconscious, which governs thoughts. Perhaps they are both right. To direct information from our world to the world of forms, we need to process it intellectually. At night, our consciousness activates intellectual algorithms to transform our earthly experience into the language of symbols and directs this information to the familiar world of abstract forms. We discussed this in chapter 11 Dreams and coincidences.

Since symbols are the best representation of abstract forms, they allow direct communication with the world of the ideal. They may also be used as a communication protocol that connects us to that world. This is what distinguishes a human from a computer, for which the language of symbols and connection to the world of the ideal are inaccessible.

We already discussed in chapter 3.9 Idealism and information the model of British theoretical physicist Roger Penrose, who popularized the concept of three worlds: the world of Platonic forms, the Physical World, and the Mental World (Penrose 2007). Figure. 3.9 illustrates their interaction. Symbols are the common part of all three worlds. They are utilized by the Mental world. Simultaneously, they have projections in the Physical world in the form of images, drawings, and objects. They are most suitable for interaction with the world of Platonic forms-meanings.

Let's try to refine Penrose's diagram and understand the role of symbols. Figure 16.4 illustrates how the Platonic world of forms-ideas defines our world. The part of the form that penetrates our world manifests as consciousness in the case of complex living organisms. The more complex the organism, the stronger the penetration of this form into our material world. Inanimate objects are merely projections of abstract "blueprints" onto the atoms and inorganic molecules of our world. They bring matter into the necessary shape and behavior. The form determines structures and processes for any material objects but does not create consciousness. For example, a computer is simply an object. It possesses elements of the simplest intelligence for calculations, but it is just a projection of an abstract form without penetration into matter, so the computer does not and will not have consciousness. Consciousness is part of the form that interacts with our world through intellect and sensory perception. And symbols are how the form is represented in our world. They are simultaneously the language of consciousness and a language for interacting with the world of forms. Symbols most accurately reflect the form-idea and are also capable of describing our world.

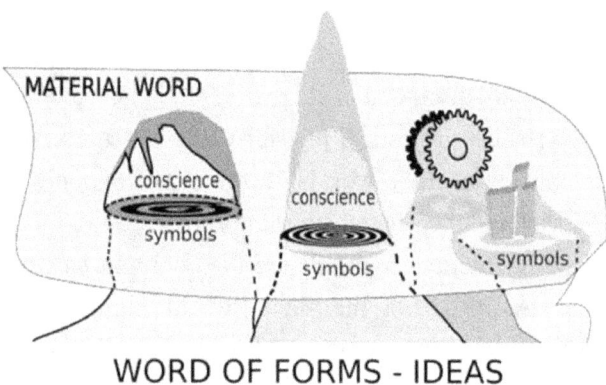

WORD OF FORMS - IDEAS

Figure 16.4. *This figure shows how the world of Platonic forms and ideas define the material world. A person with their consciousness is shown in the center. Consciousness is a part of the form. The mind includes intellect and consciousness, but the intellect belongs to the material world. All things are connected to forms, but they do not possess consciousness. Symbols are the language of forms, which are interpreted by our consciousness before being processed by the senses and intellect.*

    Numbers and mathematical formulas can also act as symbols that directly communicate with consciousness, and thus, with the forms of ideas. I often spoke about the number "6". But in all our examples, this number was used as a symbol, with which human consciousness works directly and influences the ideal form. This is because the symbol of this number has a reflection in the corresponding abstract form, which exists outside space and time, and which can influence the events and objects of our world. Such influence is not subject to time and distance.

Practically all events of synchronicity discussed in this book can be explained as manifestations of abstract forms from the ideal world as a result of the transformation of symbols to this world, and vice versa. There is only one problem — it is not always clear how symbolic-sign information corresponds to abstract forms. Perhaps such a connection is ambiguous indeed.

Let's consider the example with Kaiser Wilhelm (see chapter 6.6 The Kaiser and the War). Perhaps, in the world of abstract ideas, he corresponds to a form that is a subclass of the form of a human. Of course, these meaningful forms are exceptionally complex and contain ideas of the human body, its parts, clothing, and so on. Perhaps in that world, there are some shorter designations for these forms. Such as "6" — for an abstract person, "666"—for something related to the devil. There is also the form-idea of the alphabet. When people during World War I began to search for the "antichrist" to explain their grief, they linked the symbols "6" and "666" with the alphabet, inventing a simple algorithm that produces the number 666 from 6. Notice the peculiarity of this algorithm. We're not simply using the mathematical addition of 6 with the alphabetical order of each letter. "6" was used as a symbol, not as a number. Symbolism was clearly at play here in crafting such an algorithm! But the algorithm itself was very simple. Its sole purpose was to select the letters so that they formed a certain word — "Kaiser", which may have even been unknown at the time. This combination of letters became associated with "6" (human) and "666" (antichrist). In 1915 (and before that time), Emperor Wilhelm (and his father) was called by some other title. The desire of many people was to associate the symbol "Kaiser" with the emperor, and therefore — with Wilhelm, upon whom the lives of tens of millions of people depended.

Thus, the new word "Kaiser" transformed into a new symbol using the (already known) symbols "6" and "666". This new symbol became attached to the idea-form of Emperor Wilhelm in the minds of many people. It is difficult to say exactly when this happened. But when it did, this new form "Kaiser" was assigned to the form representing the German Emperor in the world of ideas. Unlike us, there is no concept of time for form-meanings. Therefore, the emperor form with the attached "Kaiser" form instantly edited historical information (relative to the year 1915). The word "Kaiser" became widely used to denote the emperor. Naturally, revisions of historical information were subjected to all records where the word "Kaiser" was not used. This is because everything in the material world is just actions of ideal forms "dressed" in molecular substances. Now the circle closed, bringing into balance symbols, forms, and people's expectations. As a result of this, it was easy to transform the word "Kaiser" into "666" using a simple algorithm. This does not mean that the past relative to 1915 changed. The past is already gone, it does not exist. It is neither measurable nor observable, so changing information about the past does not lead to any anomalies in the course of history.

It should be noted that some sequence of events in the world of forms must still occur. Perhaps forms have some analogue of time. However, their arrow of time does not correspond to ours. What had a sequence there occurs simultaneously for us.

Let's now look at a more complex case, described in chapter 6.1 Abraham Lincoln and John Kennedy. Here, multiple symbols were being produced. It is possible that people wanted to link the fates of these two politicians. Symbols (or forms) were created, influencing idea-forms (or symbols), which changed the information about the nature of events and names. This continued until we began to see statistically impossible coincidences in their biographies. Likely, there were multiple instances of the editing of history up until the moment of Kennedy's assassination.

When we see similarities in various details of certain global catastrophes, it may reflect the existence of some kind of proto-form. It is maximally abstract. Or, in other words, it serves as a "template" for a group of more specific forms. In the case of Lincoln and Kennedy, their two forms utilized features of this original proto-form. These two forms contained all the details of individual events. However, on a global scale, they followed the characteristics of a more fundamental proto-form.

For programmers, these analogies should be perfectly clear. In programming, there is a concept known as an abstract class. This is analogous to a proto-form. Such an abstract template describes a certain general state and behavior that future classes will possess. It is said that abstract classes are intended to create basic functionality for descendant classes. The derived classes then implement all of the abstract functions.

Here, perhaps, is the main idea that can clarify how incredible coincidences occur. In a similar way, all events of synchronicity in chapter 6 Examples of significant coincidences can be explained. Furthermore, it elucidates the incredible beauty of mathematics and physics described in chapters 8 Waves of quantum mechanics probabilities and 9 Coincidences in numbers.

It is precisely the expectations of the collective consciousness, interacting with the idea-forms on which our world is built upon, that lead to changes that subsequently appear to us as incredible coincidences. If random chance cannot be a good explanation due to the very low probability, then there must be some profound reason for such coincidences. Perhaps it is precisely the deep connection of people and things with their forms and proto-forms in the abstract world of ideas that provides the interaction, which we perceive as unrelated random coincidences.

Our hypothesis, though it may seem fantastical, is consistent. It is capable of explaining the coincidences described in this book. This assumption is no worse than the hypothesis of an infinite number of universes. Let us recall that in the multiverse model, an observer continuously "flows" from one universe to another, depending on the results of measurements of microscopic particles. This is used to interpret quantum mechanics, as we discussed in chapter 8.4 Problems of interpretations. In our hypothesis, there is only one universe. But in our Universe, information about the past "elastically" changes, aligning it with the present during rare moments of the greatest life-changing stresses of intangible essence of the human mind.

16.5 Past, present and future

For us, there is only "now". This concept is an artifact of our consciousness. It's the moment when Platonic abstract forms directly interact with matter. The present moment can be imagined as a point moving along the arrow of time, from the past into the future.

As we move through time, we leave the past behind. It disappears for us. It cannot be observed or measured. The past leaves faint silhouettes in our memory and in our reality. Fading in this world, the past becomes transformed notions of the world of ideas and forms. We "fill" these meaningful forms with detailed information from our life experience. For us, existing at the moment called "now", the past becomes flexible because our consciousness is in direct contact with the timeless world of ideas. The past and the present are connected through consciousness. This leads to retrocausality and phenomena of unusual coincidences, as described in this book.

We study the past through the traces it leaves behind. These are photographs, films, books and people's testimonies. But do we have a method of objective control that can establish that the past is unchangeable? Often, while examining old photographs, one can catch oneself thinking that what you see looks somewhat different compared to how we remember it. For materialism, all these things left to us for the present moment are objective evidence of the past. But material things are just shadows or projections of true abstract forms of meanings, where time does not exist. Any change in these forms instantaneously alters such shadows.

Is the past itself objective and immutable? Is it possible to design an experiment that will accurately prove that everything that happened to us in the past was exactly as it was, and not slightly different? I don't know the answer to this question.

But what about the future? For us, it does not yet exist. In this book, we have not found strong evidence that the future is accessible to us in all its details. However, there are interesting observations described in chapter 10 Premonitions of the future. If taken as evidence, it can be assumed that rough sketches, abstract templates and mathematical models of the future may exist in the Platonic world of forms. Some predictors may perceive them. But these templates are not filled with detailed information. As we move along the arrow of time, we give them content from the experiences gained in this Universe.

17 Free will

If the immaterial reality of ideas and meanings is God, then one might ask — why is it so hard to find traces of his presence in our world? Those who believe in him think that God has revealed himself to people through religion. He has made himself known through intermediaries and sacred texts. Yet, his traces in nature are not obvious to skeptics. They can always construct a line of arguments "proving" that there is no God. Disputes between believers and non-believers have lasted for thousands of years and continue unabated.

Wouldn't it be easier for God to provide more apparent evidence of the divine origin of the Universe and mankind, so that there would be no doubts about his presence? If God chose to announce himself through ancient texts and saints, as religions believe, why not remind us of himself from time to time through miracles that are easy to detect and document in scientific literature, so we can teach about his definite existence in schools and colleges?

Continuing this thought, wouldn't it be wiser for God to write the Bible more clearly and accurately, eliminating any ambiguity in its messages? Why do sacred texts contain cruel and evil passages at all? Furthermore, why not remove all suffering from the world? Wouldn't it be wiser for God to pay more attention to prayers, given that the outcomes of such communication are often unclear? Wouldn't it be simpler if God directly answered prayers and fulfilled requests? In terms of synchronicity, couldn't God increase their frequency to eliminate the need for ambiguous calculations and statistical methods like those we discussed in chapter 6 Examples of significant coincidences. If God exists, why does he hide from people?

What would the world look like if there were no such uncertainties? For believers, this ambiguity in the traces of creation is a condition for the fullness of this world. Indeed, it is uncertainties that make our world diverse, add color to it, and allow the exercise of free will. There are enough signs and hints in this world to understand the grand plan, for those who want to see them. The only thing that may prevent some people from interpreting such signs is their choice guided by free will.

Perhaps the main point might be this: Being free from constant prompts from above, without expectations of close surveillance by an all-powerful divine force, allows us to discover something new and unexpected that is of great interest to the reality of meanings on which the empirical world rests. To confuse and make the situation unclear is one way to ensure freedom of will in decision-making. Many philosophers agree with this position. *"I assume simply that the creator might deliberately leave certain matters to chance. Indeed, such seems to be required by the very existence of free will."* (Sanger 2024).

If there is solid proof that a supreme deity exists, it would immediately take away some of our independence. As soon as we know about the existence of a divine plan guided by a higher being, our actions would become predetermined and robotic. It would be foolish to do anything to achieve specific goals — it must simply be enough to pray. I think we would also not be able to set specific goals for ourselves. The only goal of all people would be to please the omnipotent entity to return to his kingdom. If we knew for sure that there is a supernatural being, then the whole purpose of humanity would be reduced to pure loyalty. We would become robots, doing what is expected of us to avoid possible punishment and criticism from the God upon which our entire existence now and after physical death completely depends.

The existence of different religions is one of the necessary conditions for the realization of God's plan. By giving knowledge about himself to different cultures at different times, he ensured necessary anonymity. Naturally, he was understood differently depending on cultural and historical contexts. Atheists use various contradictions between religions to argue for the absence of God. For example, as long as Christians believe that representatives of other beliefs do not deserve what is due to them after death, contradictions will always exist. And this also applies to other teachings that position themselves above other religions.

Let's consider this situation from another perspective: If God revealed himself to different peoples at different times, and all his revelations completely coincided in all small details, we would understand that there is a single, very specific, God. We will discover his will, and this eternal question will finally be resolved. Consequently, "races" would begin to strive for maximum closeness with him, forgetting why we are here, with the given to us freedom.

You can often hear that we have no free will at all. It is a product of our imagination. Everything we do is strictly determined by the laws of nature, cause and effect, our upbringing and surrounding environment. All of this, in turn, also originated from strictly defined chains of causes and effects. Such a view of the world leads to fatalism and forces one to resign oneself to fate and to surrender to the flow of the events presented to you.

There is indeed some predetermination of events and actions. But this concerns the work of our intellect. However, it is consciousness that initiates the algorithms of this computational part of our mind and makes the final decisions. If the intellect comes to the most optimal answer to external circumstances through calculations, then the role of consciousness is to decide whether to follow this decision or not. Consciousness does not belong to matter. It is a part of timeless reality where causal relationships, as we understand them, do not exist. We do not know how the intuitive feeling of being and self-awareness "negotiates" with the intellect to perform specific actions.

We must not forget that in quantum physics, the outcome of a measurement is not predetermined in advance, even if we know the state of the system well. Uncertainty and randomness are inherent properties of our world. We can only know the tendency of systems of particles to manifest one result or another, as we discussed in chapter

8 Waves of quantum mechanics probabilities. Perhaps, this is a consequence of the inseparable connection of the microworld with the reality in which our consciousness draws the strength for our existence.

I have often said in this book that one of the main tasks of our existence is to gain life experience. At least, such an assumption seems quite logical. If we are here, and we are incredibly complex biological organisms with self-awareness and intelligence, then there must be a purpose. The simplest assumption is that our very existence is the purpose. If so, there is only one immaterial product that can represent value to the world of abstract ideas. This is our life experience and the knowledge about how this experience transforms the soul.

Many of us likely made dollhouses in childhood, aiming to imagine what it would be like to live in them. I remember building a toy house out of Lego bricks during my childhood. I called it a sauna because I added small rubber tubes throughout this house. Then, I ran hot water through these tubes to raise the temperature inside this house, like in a real sauna. I even managed to measure the temperature by sticking a thermometer into this toy house. However, my biggest wish at that moment was to be inside that small house and experience how great it would feel. I knew every detail of my constructed structure, but I could not possibly imagine how I would feel being inside of it. It is not a leap of imagination to think that God might try to understand his creation exactly like this, i.e. through man.

The experience we gain in this world is information from within this Universe. It will be sent back, enriching the world of meanings on which everything is based. In general, this is the essence of life. This information should be collected by free-thinking people in an independent search, not by robots who constantly look to the sky for instructions.

This book shows that this world is not a collection of material objects where randomness rules. However, it cannot be considered as a proof of the existence of a specific God in a particular religious tradition. Our calculations in chapter 6.9 Nothing is accidental demonstrate that unreasonable coincidences are not consistent with the naturalistic patterns of randomness of the material world. Coincidences are manifestations of something timeless affecting this world. The incredible logic and beauty of abstract mathematical ideas, which seamlessly fit into the precise description of nature, points to the same line of thinking. But all of this leaves enough room for doubt, as I used some assumptions that may not be too obvious. Nevertheless, this is another step towards understanding this world and another reason to believe in God, for those who feel this belief within themselves.

17.1 Suffering

Suffering is the price for the right to have free will (Gooding and Lennox 2018). By becoming humans, we "sign a contract" which includes a clause called "suffering". It is the price for a ticket to enter the show.

In fact, suffering is a direct and logical consequence of being in a world of molecules. Needless to say, our bodies are constructed from them. Matter and energy obey natural laws and change over time. The increase in entropy or chaos in the Universe (see chapter 3.1 Entropy) is inevitable and correlates with the arrow of time. Any created structure must gradually degrade. The energy and atoms of the Universe, moving towards chaos, have existed since the beginning of this world. But we are not here forever. Living organisms can resist chaos for some time, using components from the environment to maintain their biological functions and embedded information. But still, we perish from the external "pressure" of that chaos.

The gradual approach of our molecular shell to the point of demise causes suffering. It is often said that if there is a loving God, he could have arranged everything differently. Then one might ask — how? You won't find a sensible answer. To create us in the form of eternal living mechanisms? To build a special universe which does not gradually deteriorate into chaos? But what about time? Another option would be to instantly destroy our aging bodies and erase the memory of living relatives to avoid suffering? This must be a very strange world!

I think we will never find a reasonable alternative to a world with suffering. Any mental construct of a perfect world without inconveniences, sadness, despair and passion, in a universe of indifferent atoms, would end in monstrous failure. You would end up with a world of eternally living and happy machines without emotions and meaning of their existence. According to all logical conclusions, our Universe looks exactly as it should if there were a God who created the existing order of things.

According to widespread esoteric knowledge, we reside in what is called the "dense world". Although this concept has been known for a long time, it was widely popularized in the 19th century (Blavatsky 1888). Unlike the subtle or astral worlds, the dense world consists of matter, which is also used to construct our bodies. This creates limitations and conditions under which inconveniences and suffering for biological organisms are a direct logical consequence of the material environment. Objects (including our bodies) cannot pass through each other in the dense world. This automatically imposes spatial limitations and leads to conflicts.

Here is the simplest example: If someone is in your way and you can't get around them, you need to somehow negotiate a passage. You can't just walk through them as if you were in the ethereal world, where there is no matter, made up exclusively of information and ideas. Therefore, the one who blocks the passage must make concessions. It is this moment of mutual agreement in the physical world that can be the source of various discomforts and conflicts. And therefore — suffering.

Or here is another example. Our body, like any article of clothing, requires care and maintenance. But the resources needed to maintain the biological shell are not limitless. The lack of such resources threatens your mission in this world. We must compete with others for the right to maintain our body in this environment for as long as possible. This again becomes a source of conflict, inconvenience and, therefore, suffering. As a result, suffering is a direct necessity accompanying our mission in this dense world. Could God have made resources unlimited? But how? After all, we live on the surface of a sphere whose area is limited by the number "π".

Here is a thought experiment: Let's suppose that some higher being decided to alleviate our suffering and began to alter the world in order to minimize the number of accidents or mass killings. But where is the threshold of suffering that defines the beginning of intervention by this being attempting to change the world for the better? Imagine that a threshold for intervention in bus accidents with 20 passengers has been set. Such accidents will stop happening. However, when there are only 19 people on the buses, accidents continue to occur.

The problem with this example is the following: Scientists are clever people. Such artificial intervention would be immediately detected by them using statistics and experiments. This would immediately pose a problem for our freedom of decision-making. Once this fact, based on a scientific approach, is identified and strictly proven, a significant portion of our freedom of action will be lost. The one who decides to alter the world to reduce suffering by making such interventions in the flow of natural events will cease to be "anonymous". As a result, we will start acting in favor of this higher being and engaging in irrational actions. For example, we will start manufacturing buses equipped with up to 19 seats. Many will begin to pray around the clock, instead of dedicating themselves to learning and the mission they themselves chose, using their inner freedom.

Even such explanations are often not satisfactory. Could this divinity improve our world by preventing extreme barbarism, such as the Holocaust? Maybe this would go unnoticed by people. What if this God simply reduces suffering in some clever and intricate way? For example, people guilty of tormenting others and committing mass murders would never be born, or they would die in infancy, also without suffering.

But what if this is already happening? Maybe the worst thing that could have happened did not occur. Consider this: A nuclear holocaust has already been prevented, because such an extremely tragic scenario was 'rewritten' through external intervention before we were even aware of it. How can this be proven?

The fact is that practically all events of this world are described by some probability density, like that shown in chapter 9.4 Infinity, sphere and randomness. In such probability distributions, there are tails — that is, very extreme events. But these excessive scenarios are extremely rare. Even if the most catastrophic events are artificially removed, there would still be some other historical circumstances that would seem very tragic to us. Perhaps the Holocaust was a somewhat less catastrophic event than if Hitler had created a nuclear bomb and destroyed the rest of Europe. Maybe this more tragic event had the potential to happen but was prevented from the outside. Or, an even more fantastical scenario, this event actually occurred, but the past was edited. How can all this be proven?

I think we have no good evidence of historical editing in the case of disasters on an unimaginable scale. Perhaps, we can only detect edits of a relatively "soft" nature, such as numbers, dates and words. Such minor changes in history can be traced through unexpected coincidences, as discussed in chapter <u>5 How to crack the code</u>.

Consider one example. Suppose that the Nazis actually developed nuclear weapons before the end of World War II, and they were ready to strike the Allies. But then an amendment was made to the past, and the development was delayed. The events that could have led to the destruction of Europe were edited. Therefore, now we know history as it is. Is there any way to detect a change in this historical flow and demonstrate that there were parallel, much more devastating historical events? I don't think we can do it easily.

There were many reasons why the nuclear bomb was not developed by the Nazis. Some of these reasons look strange, some are logical. There are numerous factors, from scientific mistakes to critical brain drain caused by Nazi anti-Semitism. We do not know whether any factors were subject to change or not. We can only answer such a question if we find some strange discrepancies that are traces of such altered events. In the case of the Nazis creating a nuclear bomb, we need more information about unlikely coincidences during the time when Germans were conducting this research.

Of course, we will never know what potentially could have happened if there had been no external intervention. But this does not make our response any less convincing. All we want to say is that the argument about a cruel God allowing such excessive tragedies does not work.

Or consider another situation: Suppose God stops the formation of cancer in small children. This phenomenon could again be studied and concluded that somehow, children do not get cancer, but adults do, even though all physiological circumstances suggest that all people should get sick. The scientific method of studying such a phenomenon would again lead to the conclusion of an external influence. And again, this would violate free will and create an artificial social environment where all our actions lose their natural behavior.

One might ask: Why do diseases exist at all? God could reduce their effect on human bodies or not allow them to develop, which means we would always be perfectly healthy. However, maybe he is indeed reducing the number of the most dangerous diseases, but we know nothing about it?

If there are no diseases, how can one imagine our departure from this world? Perhaps an allegory is appropriate here: Imagine that you are wearing a suit made of substances that are in direct contact with the external material world of molecules. It is subject to aging, like everything in this world. The arrow of time, which moves matter into motion and chaos, leads to its gradual decay. The information content of the world is decreasing, with an increase of entropy (see chapter 3.1 Entropy). This suit is prone to breakdowns, and its signaling system gives us pain, warning our consciousness. We need to stay in this suit for the allotted time and, using our freedom, do what we feel is necessary. Sooner or later, we will still have to return home. To leave this world, we need to be called. How to do this without revealing one's presence? Only through natural means — through diseases and breakdowns. In our Universe, the only way to be "pulled" from this world while remaining anonymous is to allow the body to be subjected to aging or diseases. And with them comes suffering.

But there are other situations. Imagine this scenario: A child is playing a computer game and cannot stop. You called their name once, then again, and so forth. They don't pay attention to you. In this case, your only option is to take the toy away from them. The screen will go dark before their very eyes, and they will finally pay attention to you. Perhaps this is analogous to various tragic cases?

As we see, the often-discussed argument that the very presence of suffering indicates the absence of a loving God does not withstand scrutiny. Suffering is the logical consequence of the fact that our consciousness must maintain its free will for an autonomous mission in a dense environment filled with molecules and limited resources for life. Almost all religions state that Earth is one of the most challenging places to exist, but it is precisely this complexity that allows our soul to complete our mission quickly.

17.2 Moral law in your conscience

The question of the origin of moral concepts such as "good" and "bad" has been the subject of philosophical discussions for many centuries. Various philosophical and religious traditions offer different viewpoints for explanation. Some believe that moral concepts originate from divine sources, such as God or gods, while others argue that they are products of human activity, evolution, culture, or emotional reactions to events.

You often hear the following: I don't need faith in God (of any religion) to refrain from murder, rape, theft, or assault. Suggesting that only the fear of punishment from God makes me moral can even be insulting.

But here's the main question: How do we know that our brain isn't "programmed" from birth to distinguish good from evil? What if this programming is precisely invented by God? One can disbelieve in God, but that doesn't free us from the simple fact that, delving into consciousness, we will see a morality that is given to us from outside. *"Good is nothing but the sum of the moral qualities of God."* (Old Testament, Exodus 34:5-7). What if, like mathematics, morality is real? Our brain simply discovers it, like logic, abstract formulas and numbers. Many philosophers argue that there are objective moral truths that exist independently of human beliefs or emotions.

Indeed, even chimpanzees have all the prerequisites for morality. They can exhibit compassion, altruism, fairness and even a sense of guilt. In humans, basic moral settings are present from birth. Infant studies show (Kanakogi 2022) that morality is likely ingrained in our brains from infancy. Over a lifetime, people can modify and supplement their moral concepts.

The concept of objective morality, similar to mathematical or physical truths, is also the subject of philosophical debates. Some philosophers assert that objective moral truths exist independently of human beliefs or emotions, while others argue that moral values are ultimately subjective and culturally relative.

Of course, this does not prove that God "embedded" morality in us from birth. Opponents, usually atheists or agnostics, would say that the primary source of morality is the evolution of society. In the distant past, evolution compelled people to cooperate. We had to live in large social groups, which improved our ability to get along and interact with each other. These social skills were passed down from generation to generation. They were recorded in our genes and became our instincts. Later, the cultural environment and upbringing further complicated and enhanced moral concepts.

There could be other mechanisms. For example, to lie, you need to know that there is truth. But constantly lying, while knowing there is truth, cannot lead to a stable psyche (Dawkins 2008).

For those who believe, even such a theory of the natural origin of morality cannot refute its divine design. The process of forming morality through evolution could have been intended from the very beginning. If God exists, then he does not need to "program" our brains or create various ways to recognize morality as an objective law independent of humans. He knew from the start that as long as there are groups of conscious beings interacting and trying to survive in a hostile environment, cooperation among them, and thus morality, would arise automatically.

But how can we determine whether morality is a natural process, or if, after all, God created consciousness in such a way that it could recognize good and evil as objective categories external to humans? People who see morality as a natural phenomenon would say this: The burden of proof lies on the side of those who propose a force beyond societal evolution and natural selection. I think the burden of proof rests on both opposing sides. On those who believe that good and evil are merely products of evolution, and on those who believe that these are moral categories set from the outside. This makes the problem of morality symmetrical for these opposing sides.

The unlikely coincidental events due to synchronicity described in this book appear quite indifferent to good or evil. They simply occur. We have only been able to show that there are some phenomena that depend on the strong emotional state of people, not describable in a naturalistic way.

I think that the proof of the objectivity of morality should be found elsewhere. There are numerous studies (Atwater 2007) indicating that the overwhelming majority of people with near-death experiences can confidently say that goodness and love exist beyond this world. See chapter 15.3 Consciousness outside the body. After experiencing a near-death experience, they become more spiritual, they develop greater compassion for others, and overall, they become more moral. I see no reason why a near-death experience would somehow enhance a morality instinct acquired through human evolution, while at the same time not enhancing their other instincts, which are much more vital for human survival.

17.3 Time allotted to us

Everyone decides for themselves when it is time to leave. Our bodies, interacting with matter, wear out and break down. This has a significant impact on our ability to fulfill our mission. If, due to some accidents, our mission becomes unsuccessful, we may leave earlier. There are also situations when we cannot make such a decision on our own.

However, there remains one interesting question: What about human civilization? Imagine that someday, scientific progress reaches such a state of development that the "hypothesis of God" turns into the "theory of God". God will be proven by precise scientific methods and studied in schools. For example, experiences during clinical death will be so well studied that there will be no doubt about the existence of a higher being. Naturally, everyone will eventually want to leave this world. Or people will start spending most of their lives to please God to get a comfortable life beyond this realm.

Here arises a very serious philosophical question: What is the purpose of a world where everyone knows for certain that God exists?

As we said, it is precisely the uncertainty of the existence of the Almighty that makes us free. Of course, such uncertainty does not exist for people who already believe in God and practice some system of religion. But if the evidence is absolutely precise, then religion and faith in God will lose their meaning. Irrefutable knowledge about God will take away much of our freedom and make us robots, doing everything possible to please him.

In such a situation, a world populated by people without free choice would not be very purposeful. People would not be able to effectively gain experience and make discoveries knowing that the main task is to prepare themselves for the world of God and to "slip away" from this reality as soon as possible or, at least, to minimize the number of actions that would displease the creator.

It is possible that civilizations, where God is a subject of exact sciences, and where everyone finds themselves in constant communication with him, pleasing him, praising him and his creations, begging him for something, and waiting for heavenly grace after death, would be quickly halted. It would have to be "rebooted" to the very beginning, to the point where the concept of God becomes uncertain again. And people would again immerse themselves in the environment of nature and their inner freedom, in a world where there are choices.

If this is really so, it might explain the Fermi paradox — the absence of alien messages and visible traces of extraterrestrial civilizations, which should have already come into existence in our galaxy, considering the significant age of this Universe. Highly developed civilizations that managed to uncover the secret of God either ceased to exist or were "rebooted" back into a primitive communal system. Or they destroyed themselves as a result of wars and dangerous technologies. This, perhaps, was planned according to the meaning of the social laws, which also found their place during creation.

Can books, like this one, hasten the moment when the existence of God becomes provable? I do not think so. Firstly, there will always be uncertainties that can be used to refute the idea of creation. In this book, I only showed that it is possible that the reality of this material world with cause-and-effect features is questionable. But the fact that these phenomena are associated with the presence of God, as understood in world's religions, will remain a matter of faith for many.

Secondly, it's possible that books similar to this one may not be entirely accessible to many readers. For such individuals, a specific religion may be of greater interest, as they operate within more understandable and tangible categories. Religion appeals to tradition, culture and emotions, but less to intellect and logic. And so, religion will always be subject to criticism from agnostics and religions of other traditions. This will sustain uncertainty, and our world will continue to exist in a balance between believers and non-believers. This is how it should be, if there is a rational design by the creator, leaving us freedom in decision-making.

18 Hypothesis of God

In this chapter, we will consider the main arguments and criticisms of two contrasting positions: whether God exists or not. We will not discuss issues related to religion. They involve factors of understanding God in the historical, social and cultural environment of a specific society. If we assume the presence of God, then a large number of religions exist only because the spiritual (ideal) world of God was understood by people in different ways at different times. In the past, these revelations happened depending on historical conditions and cultural experiences. If God had revealed himself to all peoples on Earth in the same form and with a single scripture, it would imme-diately have lifted the veil of anonymity and uncertainty in his exist-ence. This would have negated much of our free will, as we discussed in chapter 17 Free will.

18.1 There is no God

The main argument of atheists is quite simple: There is no God because there is not the slightest substantiated evidence. Every-thing we see around us is perfectly consistent with the assumption that matter and energy are sufficient for the world we observe.

We have already discussed that this is not the case. The pres-ence of information is already a sign (but not proof!) of a deviation from the materialistic concept of the world. Information cannot be cre-ated by matter alone. This last statement cannot be proven. But many can believe in it, based on millennia of observations of nature by peo-ple. See chapter 3.5 Life and information.

The concept of God suggests a specific purpose for us. To ful-fill such a purpose, the anonymity of God and the lack of concrete evidence of his existence are essential (see chapter 17 Free will). In my opinion, any scientifically valid proof of God would be incompat-ible with the existence of intelligent life. As a consequence, there would be no humans who could question the existence of God.

One of the most popular opinions was well expressed in the book "The God Delusion" by Richard Dawkins, a British biologist and science popularizer (Dawkins 2008). At the end of one of the chapters, he concludes:

"In the case of a man-made artifact such as a watch, the designer re-ally was an intelligent engineer. It is tempting to apply the same logic

to an eye or wing, a spider or a person... The temptation is a false one, because the designer hypothesis immediately raises the larger problem of who designed the designer. The whole problem started out with was the problem of explaining statistical improbability. It is obviously no solution to postulate something even more improbable... We need a `crane`, not `skyhook` ... The most ingenious and powerful crane so far discovered is Darwinian evolution by natural selection... Some kind of multiverse theory could in principle be do for physics... God almost certainly does not exist."

In this book, perhaps the strongest argument concerns the natural sciences. We will not consider the moral arguments that are presented later in his book.

The first thing that strikes one in such reasoning is that the significant part of the problem of God is reduced to the Theory of Evolution (in its modern understanding). As we have discussed, the main issue lies in the emergence of initial information, when individual molecules came together in large groups and "came to life" creating instructions in DNA and the first primitive cells. See chapter 3.5 Life and information. This critique is well expressed in several books by Stephen Meyer (Meyer 2009) (Meyer 2021). We know of no natural process in the world where information is created from matter. The theory of evolution does not work at the level of simple biological structures emerging from molecules.

Regarding the theory of many universes, we have discussed it earlier. Very few scientists will explain one unknown, that is, our Universe, with an infinite number of unknowns, being an infinite number of universes, the reasons for whose existence are also unknown. Such reasoning leads nowhere and is not scientific.

Further, Dawkins' criticism primarily pertains to Christian religion and the Bible. In our book, we did not delve into religious doctrines. The problem of God is a much broader issue involving faith, culture, the interpretation of science, mathematics, philosophy, and, as we have shown, even statistics.

I will no longer touch on questions such as whether the Theory of Evolution is a sufficient framework to describe all of the details of historical events (see chapter 3.6 Theory of evolution). Or, whether morality, with its concepts of good and evil, can arise in the process of societal evolution without an absolute external standard. For me, these are simply hypotheses, models, or qualitative explanations that attempt to describe the distant past. Hundreds of millions of years of historical development of a rich information-based environment can easily challenge any naturalistic model to describe all the details of the world as we observe it now. Anything can happen over such an unimaginable time. Any random occurrence, beyond even our ability to conceive. We lack historical data for quantitative description. Therefore, we can only construct hypotheses and models to explain how such quantitative changes lead to qualitative leaps in information, even if we use the most accurate theory proven in laboratories.

Here's a simple example from high-energy physics. In this field, there are many beautiful and non-contradictory models and hypotheses. They are developed by a vast number of theoretical physicists. They are vigorously discussed at crowded conferences. Attempts are made to find or refute them. However, there is one problem: None of these models have yet withstood experimental testing. Each time experimental data refutes such hypothetical models, it turns out that they have so many tuning parameters that they easily come out "dry". And they need to be tested again, but under different conditions.

Materialism and atheism build a picture of the world without invoking the supernatural, yet they use the same philosophical concepts and logic that is found in idealism and religion. There is no answer to the question of where logic and natural laws come from within the worldview of atheism. Its limitation lies precisely in denial, creating a system of belief that limits the capacity for fantasy and exploration. Here is what the Russian scientist and publicist Nikolay Karyshev (1855 – 1905) wrote:

"The purpose of science is to study and investigate everything it does not understand. Atheism, however, does the exact opposite regarding questions about God, the afterlife, the human soul, etc.: it simply denies, without investigating, studying, or refuting, but simply and unconditionally denies. This is the lowest understanding of the nature of things, if one can even call atheism an understanding; it is the absence of any desire to understand the nature of things". (Карышев 1895).

Nevertheless, the position of atheism plays a tremendous role. It is inherently laid out in creation and provides the necessary balance of forces between the spiritual and the material. Atheism is a source of skepticism, without which the pursuit of truth is impossible. Creating a planet of people who worship, praise, and petition a deity cannot be the purpose of creation, if such a task truly existed. This has long been noted by materialistically-oriented critics of religion. Uncertainty gives us the freedom to choose (see chapter 17 Free will). And to find the truth, one must embark on a journey of knowledge.

The fact that our bodies and brains grow from a few cells and do not "pop-up" in their final form into existence suggests that we do not have an initial memory of God (or other things). This knowledge must emerge in the process of biological and spiritual growth, observing and analyzing the world around us. And for any development, the presence of choice is necessary.

Those who hold the opinion that there is no God find support in this statement: If God created everything around us, then who created God? If he exists and is incredibly complex, then there must be a first cause for his existence. Some Super-God who created "our" God. And so on. We already know that using infinities to explain anything never leads to a reasonable answer. This way of thinking is deeply contradictory. The British philosopher Bertrand Russell (1872 – 1970) even used this contradiction to justify his disbelief. We will discuss this problem in the next chapter.

18.2 There is a God

When speaking of a God, the first assumption we make is that God is beyond matter, space and time, which are his creations. According to all religious teachings, God cannot be created. If he had been created, then he would not be God. He is the first cause. Aristotle also said that by tracing the chain of causes, we inevitably come to the first cause — or God. If we assume that the chain of causes never ends and goes into infinity, then you will not be able to make a single reasonable statement. Any answer will be meaningless.

Previously, we said that any complex and functional mechanism requires information created from the outside by a mind. For God, this is not the case. He is the first cause of information. He is both its creator and consumer.

If you think that in the distant future, science will be able to explain how information arises from atoms without the need for conscious activity, you are mistaken in your judgment. People often use an example from ancient times. The discussion goes like this: You see, at that time people didn't know what thunder was. The phenomenon sounded like the rumble of drums, so they invented gods to explain this noise. Later, science figured out that it's just a physical phenomenon. Hence, God must be a "god of the gaps" in our knowledge.

It is immediately worth noting that such examples are not directly relevant to the issue at hand. In ancient times, people attributed even the simplest phenomena to deities. However, the presence of such precedents in history speaks little to their relevance in the present day. We have long moved away from the idea of associating natural phenomena with higher powers. Science effectively explains natural phenomena through other natural occurrences. The issue concerning God lies elsewhere: Where did these chains of phenomena originate? Where is the source of natural laws? From where did the functional complexity and the associated information required to create and operate this complexity arise?

Here is a type of question from the same category: If you see a vacuum flask with tea inside flying through space, can you say that you just need to wait long enough, and science will explain it? A vacuum flask is an example of functional complexity. The only objection you might hear is the following: We know that vacuum flasks can be made by people because we have many examples of people making similar bottles. But is that important? Can we apply categories in our judgments that we have never encountered in our lives? The point is not whether people have made such items before or not. Our intellect and consciousness are capable of identifying functional complexity, regardless of past experience.

The fact is that there are billions of people living on this planet, and not a single one of them has seen functional complexity arise by itself. Science works in exactly this way — using observations and experimentation. And here the conclusion is quite clear — the natural laws of nature are not capable of this. We don't need to wait "long enough" to explain the complexity of structures interacting with an informational code. It simply cannot happen. Even if you believe that it takes billions of years of nature's "attempts" to create meaningful information encoded on a material medium, you are mistaken. Those same billions of years can also destroy any sign of a randomly arisen complex structure.

This book has demonstrated that the presence of meaningful information exists even now. The probabilities of unusual events clearly indicate that they do not occur by chance. This timeless "force" is capable of distorting events, and perhaps, it adjusts the past in such a way that the present appears as it does now. It makes decisions — therefore, it is capable of this and, consequently, is intelligent. It is connected to the minds of people and manifests through them. This phenomenon, which exists outside of our time arrow, influences both the past and the present of our world.

This incredible reality, beyond the existing one, reveals itself via synchronicity, linking human conscience with matter, event sequences, time periods, and various numbers. Mathematical regularities discovered in numbers, fundamental constants, and equations favor the decimal numeral system. All such phenomena point to humanity: The decimal counting system, which is used in all such coincidences, emerged due to the structure of the human hand. Additionally, it also uses the sexagesimal system for time measurements, given to us by ancient civilizations.

Our brain is also given preference. We are capable of creating abstract constructs that find their place in describing this world. Humans can operate with infinities and yield finite expressions. We can even create mathematical abstractions, such as imaginary numbers and symmetry groups, that are completely meaningless in the world of things we deal with every day. However, as centuries pass, all such concepts are found to be absolutely necessary for describing the microcosm.

Unlike sciences, which answer questions about how the world is structured, how phenomena are interconnected, and what laws govern it, the hypothesis of God answers the most fundamental question: Who or what is the ultimate "source" of everything we see, and why the laws according to which this eternal motion occurs have arisen? In billiards terminology, this is a question of who (or what) took the cue and struck the first ball, setting all the balls in motion. After all, all motion occurs according to laws that were established in advance. It happens on a table that was created for these purposes before anything starts to occur. You can observe a game of billiards and ask scientific questions such as "how do collisions occur", "how to calculate the trajectory of a ball" or "which ball will bounce off another and when". But you can also ask questions like "why?" or "for what purpose?" Why does the billiard ball exist? Perhaps the billiard ball and even the table itself exist solely because the game exists. But this game needs to be designed.

Thomas Aquinas (1225 – 1274), a philosopher and theologian, expressed this idea, although similar reasoning can be found in Aristotle: Things are constantly "moving" (moving, changing, emerging, and being destroyed). Thus, they strive to realize a multitude of diverse potentials. Therefore, an "eternal immovable Prime Mover" must exist. This Prime Mover must be outside matter, which they set in motion. Since they did this, they were capable of decision-making. And thus, they were intelligent. Thomas Aquinas presented five proofs of God, and the proof of motion was one of them.

Another proof is about the cause of all things. All cause-and-effect relationships are strictly ordered and arranged in chains. If there are such chains of causes and effects, then there must exist an Uncaused Cause. This Uncaused Cause is God.

The third proof is as follows: The world consists of things that can either exist or not exist (be destroyed). These things are not necessary. Therefore, they are of a random nature. One can imagine a moment when all things ceased to exist. But if such a moment were possible, the world would have long since disappeared. Therefore, there must be something necessary, having its cause within itself and never disappearing.

The fourth proof is about the existence of absolute perfection. All things possess various degrees of perfection. But if there is a spectrum of degrees of perfection, there must be a maximum limit to all perfections, therefore, there is a cause of such maximum perfection. Such a being is God.

The fifth proof is about purposefulness. Everything in the world is harmonious and purposeful. But if there is purposefulness, there must also be a highest purpose — a mind responsible for order and well-being in the world.

These proofs may be subject to doubt by those who believe that the Universe contains infinite motion and infinite chains of causes and effects, where there is no purpose. Perfection and purposefulness are subjective and entirely random characteristics. If you can envision all this, your worldview leans towards materialism and atheism. Or you are an agnostic, as you have no definite opinion due to a lack of evidence.

Philosopher Immanuel Kant disagreed with the classical proofs of God proposed by Thomas Aquinas, although Kant was very far from atheism. In his view, the existence of God cannot be proven by any theoretical proof. God is transcendent. He is beyond our reality and our understanding. Therefore, he cannot be comprehended. Faith is required. Nevertheless, Kant proposed his proofs based on the existence of morality and ethics. We have already discussed this issue earlier.

18.3 Scientists at a crossroads

In scientific communities, where mechanisms of cause-and-effect relationships in various processes involving matter and energy are studied, skepticism towards phenomena beyond naturalistic explanations is quite common. Indeed, researchers are dealing with exploration of naturally occurring phenomena. They formulate hypotheses and subject them to doubt in order to find the right answer. It is precisely skepticism and critical thinking that allow establishing logical chains of connections and understanding the reasons why one phenomenon follows from another. Scientists need substantial naturalistic foundations, laboratory experiments and independent observations to find possible explanations and derive laws of nature.

And here is where the most interesting part begins: It turns out that many scientists believe that a purely materialistic description of the world is insufficient to answer all the questions we encounter. A survey of members of the American Association for the Advancement of Science showed that 51% of scientists believe in God or some higher powers (Masci 2009). However, this figure is much lower than the 95% for the entire American population who profess such a belief.

I know many scientists who have come to the opinion that life is not the culmination of cause-and-effect chains of natural phenomena and random processes of nature. Often, they acknowledge the existence of God or interpret the divine power within the framework of a particular religion. But this does not mean that such personal views stop their scientific exploration of causal relationships in nature. The freedom to choose a worldview is one of the privileges given to us at birth.

Most scientists are very busy. They identify cause-and-effect relationships between phenomena, describe observations, and build experiments. Finding time to "rise above" such specific problems and look at their totality from above, to generalize and ask the most important question, is difficult. If they try to ask themselves what it all means, most often the answer will be that we do not know, or someday science will explain everything. The latter is simply a certain evasion from answering the question. Of course, many can answer it without waiting for future centuries. It seems better to try to answer the question, even if this attempt leads in the wrong direction. The wrong answer can be corrected in the future, while not attempting to answer the question usually leads nowhere.

But there is yet another component here. The God hypothesis can be considered irrational because no single experiment can prove his existence. Consequently, this perceived irrationality might call into question the rationality of these scientists in their everyday work. This can also reflect on their careers. Explaining natural phenomena by an external intention is not a scientific method. The condemnation might look like this: If you believe in this, then it means that you may not apply scientific principles in your work. What could be a worse characteristic for a researcher?

The problem with such logic is this: The belief in a designed world surrounding us and the belief in the primacy of matter represent two distinct worldviews. They are not used for solving equations, multiplying numbers, analyzing data tables or finding any dependencies. Scientific success is not related to belief or disbelief in a deity. In most cases, scientists with non-materialistic views of the world simply ask more questions about the meaning of everything that exists, and they are inclined to answer such questions.

I have already talked about the life of Augustus De Morgan in chapter 9.4 Infinity, sphere and randomness, the great Scottish mathematician who, while being the president of the London Mathematical Society, hid his interest in studying spiritualism, as it could negatively affect his colleagues' attitude towards him. Nevertheless, he was the anonymous author of the book "From Matter to Spirit: The Result of Ten Years' Experience of Spiritual Manifestations".

Here is another example that shows how people involved in natural science come to believe in God as they accumulate enough knowledge over their lifetime. Nikolay Pirogov (1810 – 1881) was one of the most famous Russian surgeons, who created several new branches of medicine, widely applied the use of plaster casts, and founded the Russian school of anesthesia. In the middle of the 19th century, all of medicine was exclusively materialistic. At that time materialism captured the imagination of many scientists. Pirogov was also a consistent materialist for almost his entire life. Having performed over 10,000 surgeries and seeing how heavily life depends on the body, it was very difficult to believe in God and that there is a spiritual principle. Pirogov noted: *"A doctor is daily convinced visually that all mental abilities are not only connected with the body, but also fully dependent on it..."*. Like many at that time, Pirogov followed Darwin's teachings. However, at the same time, he posed the question, *"What made the atoms of matter assemble into a form which is capable of autonomous existence and struggle for survival, heredity, and producing new beings similar or dissimilar to itself?"* He began to doubt whether the primordial cell *"did not contain creative thought in its ultimate purpose of creative (purposeful) predestination"*. Pirogov concluded that the human organism is simply a device made of atoms, which was necessary for the "world thought" to exist in matter.

In the last years of his life, Pirogov experienced an extraordinary spiritual enlightenment. *"For me, the existence of the Supreme Mind and Supreme Will became as necessary as my own intellectual and moral existence,"* he wrote (Пирогов 2010). He intuitively began to feel that the cause of all phenomena in the world exists in some mysterious environment, in "the boundless", which created life, time, space, and matter. He recorded in his diary these lines:

"The vital principle can be compared to light or the light ether, something unlike matter, capable of penetrating through materials impenetrable to any other matter, and at the same time imparting new properties to them".

Do you find this expression familiar? Indeed, this idea permeates our entire narrative. He envisioned the Universe as rational and the activity of the forces operating within it as purposeful and meaningful. In his view, "I" is not a product of chemical elements, but an embodiment of the general *"Universal Mind, which I imagine to be acting freely according to the same laws that are prescribed for my mind, but not constrained by our humanly conscious individuality"*. This view is very similar to that of Konstantin Tsiolkovsky (1857 – 1935), a Russian scientist who developed theoretical issues of cosmonautics. He also perceived the Cosmos as a single living organism.

All these thoughts of Pirogov became available after his death from his diary, which he wrote "exclusively for himself". He was against these thoughts being made public during his lifetime. Such views were not welcomed in the 19th century when complete determinism was the only accepted approach in medicine.

To conclude the description of Nikolay Pirogov's life path, I will mention a rather interesting fact. He died of jaw cancer in the village of Vishnya (now the city of Vinnytsia). According to his son's recollections, before the onset of Pirogov's agony, *"a lunar eclipse began, which ended immediately after the resolution (death)"*. If we had more data, we could say whether this phenomenon was a mere coincidence or a manifestation of synchronicity. It is quite possible that it was the latter, as the deaths of spiritually enlightened individuals are often accompanied by phenomena of synchronicity, at least for those who witnessed such events.

Thus, Pirogov did not consider the simultaneous acceptance of science and faith as impossible or absurd. There are many examples of famous natural scientists for whom the existence of God or nonmaterial higher powers did not contradict the scientific method. Here are just a few names:

- For Isaac Newton (1643 – 1727), the existence of God was a conclusion that could be drawn from observing Nature.
- Charles Darwin (1809 – 1882) believed that God was the supreme lawgiver and was convinced of the existence of God as the first cause. An analysis of his views (Meyer 2013) shows that if he had known about the incredible complexity of a cell, his views on the formation of life and complex organisms would have been fully theistic.
- For James Maxwell (1831 – 1879), a physicist known for his classical theory of electromagnetic radiation, science was a deeply religious pursuit.
- Michael Faraday (1791 – 1867), who contributed to the study of electromagnetism and electrochemistry, was a devout Christian.
- Dmitry Mendeleev (1834 – 1907), a Russian chemist known for discovering the periodic law of chemical elements, considered serving God the main duty of his life.

- Guglielmo Marconi (1874 – 1937), an Italian inventor known for creating the telegraph based on radio waves, was a Catholic.

- Albert Einstein (1879 – 1955) believed there is a "lawgiver" who set the laws of the Universe, and who reveals himself in the ordered harmony of the world.

- Arthur Compton (1892 – 1962), an American scientist who won the Nobel Prize in Physics in 1927 for the discovery of the Compton effect in 1923, was convinced of the existence of God and believed that the spirit of Jesus still exists in this world.

- Konstantin Tsiolkovsky (1857 – 1935), a Russian and Soviet scientist who developed theoretical issues of cosmonautics, believed in the existence of God as the creator of the world or "first cause". For Tsiolkovsky, the cosmos was a single living organism.

- Max Planck (1856 – 1949), a German theoretical physicist, founder of quantum physics and Nobel Prize laureate, believed in an omnipotent and omniscient God.

- Werner Heisenberg (1901 – 1976), a German theoretical physicist and one of the main pioneers of quantum mechanics, was a deeply religious person with a strong conviction in the Christian faith.

- Erwin Schrödinger (1887 – 1961), who achieved fundamental results in quantum theory, had a deep connection with Hinduism and Buddhism.

- Jacob Bekenstein (1947 – 2015), an Israeli-American theoretical physicist, made fundamental contributions to the foundations of black hole thermodynamics and the interrelationship between information and gravity.

This list can go on indefinitely. *"It was not by accident that the greatest thinkers of all ages were deeply religious souls"*, claimed Max

Planck (1856 – 1949), the founder of quantum mechanics. Many outstanding scientists, such as Einstein and Feynman, about whom we have spoken, were not directly religious. However, they were almost certainly spiritually enlightened and sought the cause of existence beyond matter.

Scientists should not be afraid to ask questions, the answers to which far exceed the bounds of scientific inquiry. It is precisely such questions that provide the reasons for engaging in science. Attempts to give answers that are unimaginable to others, simply relying on logic, philosophy, inner feelings and aesthetic principles, can lead to discoveries in many centuries from now. This has been shown many times in this book.

Adopting the same approach as that used in scientific endeavors, which prioritize logic, feasibility, elegance and beauty, you may choose to tackle the grand question: Where did all these laws and naturalistic principles come from? Not from themselves, surely? Any attempt to answer this question using science itself leads to a vicious circle of arguments. Perhaps you'll be ready to break free from it and turn to new possible answers that do not fit within the framework of today's natural sciences. In doing so, you can leave all these scientific principles to themselves, and treat them solely as tools somehow given to us for reverse engineering aimed at reconstructing the design of the entire project of nature.

It is believed that the overwhelming majority of Nobel laureates believe in God in a broad sense, not necessarily adhering to the tenets of religions (Shalev 2002). Among scientists, this number might be smaller. But even the description of religious views in the book "50 Nobel laureates and other great scientists who believe in God" (Dimitrov 1995) looks impressive.

Here is what Albert Einstein wrote in a letter to the Indian philosopher Rabindranath Tagore (1861 – 1941):

"But also, everyone who is seriously involved in the pursuit of science becomes convinced that some spirit is manifest in the laws of the universe, one that is vastly superior to that of man. In this way the pursuit of science leads to a religious feeling of a special sort, which is surely quite different from the religiosity of someone more naive." (January 24, 1936).

Einstein was never an atheist, although he did not support the religious postulates about the personalization of God.

A typical start of the conversation that you might hear between a non-believer and a believer goes something like this: The skeptic can say that "sooner or later, science will explain everything we observe around us. God is just a god of the gaps in our understanding of the world, it simply reflects our ignorance."

To answer such a claim, we should again point out that God, as a way to fill gaps in the description of the world, is no longer used in either philosophy or theology. The concept of God does not arise from ignorance of how this or that mechanism works, but rather from the knowledge that the nature of the questions themselves can be different. If you see the operation of a complex device, you might ask how it works. This is a question that science and engineering can answer. However, if we ask about the purpose of its existence, meaning or its cause, we may arrive at completely different types of answers. Secondly, the statement "science can explain everything" cannot be derived within science itself. You need something else, but what? We have discussed this topic before.

One way to counter the argument about God as a means to fill gaps in knowledge is to bring up the example of information and say that the very emergence of information in the material world is the best justification for the existence of God and, hence, meaning. The material world cannot create information by itself. On the contrary, information can transform the material world. No, this is not strict scientific proof. But we are not looking for one, because there should not be one if there is meaning to our life and God (see chapter 17 Free will). However, our experience tells us that everything that has functional complexity has a cause. This means it was created for something in which meaning, and information were embedded. First, there was a choice, followed by a decision, which then led to the design and creation. And any decision is the privilege of conscience.

Science, as a method to understand the workings of nature, is often mistakenly conflated with deeper philosophical questions. John Lennox, a professor of mathematics from Oxford, expressed this very aptly (Lennox 2019):

"In regard to God as an explanation in competition with scientific explanation. That is as wrong-headed as thinking that an explanation of a Ford car in terms of Henry Ford as inventor and designer competes with an explanation in terms of mechanism and law. God is not a 'God of the gaps', he is God of the whole show."

The idea that the concepts of God and science cannot coexist distorts both the process of science exploration and its results. Science, philosophy, history and spirituality are the integral parts of the human quest for truth and meaning. This thought is eloquently expressed in the book "Why Science Does Not Disprove God" (Aczel 2015) written by mathematician and science popularizer Amir Aczel. The aim of this book is not to defend the theory of God of any religion, but to defend the integrity of science as a way of understanding the world. We cannot arbitrarily manipulate the scientific method to disprove the idea of God. This concept delves into deeper questions than any naturalistic approach alone can address. It represents an endeavor to tackle profound existential inquiries by employing the full spectrum of human knowledge and understanding. Even scientists who were not religious, such as Peter Higgs (1929 – 2024), believed that science and religion are not incompatible. He said, *"I myself am not one (a believer), but perhaps it is more a matter of my family background than the fact that there are some fundamental difficulties in reconciling science and religion"*.

Many phenomena and processes of the material world, such as floods and volcanic eruptions, may not have a deliberate and rational cause. They can occur without external intervention, as they obey laws that were set from the beginning. Such natural phenomena are similar to an alarm clock ringing, which arises from a mechanism that triggers the ringing through interacting gears. However, the mechanism of the alarm clock itself, the physical laws by which it operates, and even the spring of the alarm clock were wound by the owner of the mechanism. How such phenomena manifest and are perceived, at what time and with what consequences, depends on the natural laws and the observer. From the text of this book, we already know: When circumstances lead to an emotional experience, one can discover amazing coincidences and paradoxes. They do not lend themselves to rational understanding if we assume that impersonal, inanimate matter and energy are all that define our world.

19 Epilogue

Among the articles and books on particle physics and scientific programming that I have written, this book is particularly special to me. I tried to combine a scientific approach to explain unusual phenomena with philosophy and spirituality. This book ventures into the realm of speculative hypotheses, seeking answers to questions that extend beyond our current knowledge. To understand this world, experiments and scientific proofs alone are not enough. Searching for explanations for questions unthinkable to science requires all available tools that people possess, including philosophy, tradition and religion. This book is a reflection on questions to which science has no answers.

There are no strict scientific proofs for the existence of God. If there were, this world would lose its purpose. Our universe, along with us, simply would not exist. Therefore, the lack of precise confirmations is a minor "inconvenience". We paid for it by coming into this world. Like many other books on spirituality and meaning, this book cannot definitively prove the existence of God in any specific religious context. However, according to all observations and logical conclusions that can be drawn from this book, our Universe appears exactly as it should if God exists.

I hope that for those who are seeking the meaning of life, this investigation provides sufficient grounds for the validity of the hypothesis of a God — as a creating, rational and loving reality, influencing the material world at all its scales. If my arguments are not enough to discard atheism and to adopt the belief that at the foundation of this world lies a rational reality that created this world, still interacts with it and influences the fates of people, then my only response would be this: follow your feelings and intuition. And remain a materialist. Perhaps that is your mission in this world.

God, splitting himself into billions of parts, erasing their memory at birth and giving them free will, explores himself and his creation. He skillfully veiled his presence. He seeks to understand and experience what we perceive as the Universe. Those who see no meaning in God also play a role in this spectacle of life. Perhaps, their mission is to gain experience in a world where there is no God. Simply put, he decided to test whether it is possible to justify the world he created without his participation. Did he leave enough clues of his presence to foster belief in him, while remaining undetectable by any device? Do you feel a sense of wonder about the place you are in, and a longing for your true home to which you will return? And regardless of whether the answer is positive or negative, you will still connect with him, fulfilling your purpose. He will embrace you and accept you into his arms. This book is just another "open gate" of hope for those who strictly follow logic and science. And perhaps, at some point in their lives, it will help them in unveiling the incredible realm beyond.

.

20 Additional materials

Here I present the Python codes that were used in our calculations throughout this book. Python is a high-level programming language distinguished by its efficiency and ease of use. It is widely used in the development of application software, as well as in machine learning and big data processing. One can use Python written in C or Java. In the latter case, one can perform computations using a free program called DataMelt (Chekanov 2016) that runs on any operating system. As the author of this program, I take this opportunity to recommend it to you.

There are quite a few books on how to program in Python. I hope this code is clear enough for those who already have some programming experience.

20.1 Appendix

In this code for chapter 3.1 Entropy we calculate the Shannon entropy for several lines of text. We break the text down into letters and calculate the entropy for three sentences.

```python
# Calculation of Shannon entropy from string
import math
from collections import Counter

def shannon(text):
    data=list(text) # break to a list of letters
    counts = Counter()
    for d in data:
        counts[d] += 1
    ent = 0
    probs = [float(c) / len(data) for c in counts.values()]
    for p in probs:
        if p > 0.: ent -= p * math.log(p, 2)
    return ent

txt="I LIKE SNOW"
print(txt," Entropy=",shannon(txt) )
txt="I LEKI SNOW"
print(txt," Entropy=",shannon(txt) )
txt="I LEKI SNOWMAN"
print(txt," Entropy=",shannon(txt) )
```

20.2 Appendix

In this example for chapter 3.4 About probabilities and information we calculate the probability of forming the sentence "I LIKE" by iterating through all letters of the English alphabet. The obtained probability is about 3×10^{-9} from ten successful attempts. Note that the program requires a considerable amount of calculation time. Therefore, for testing purposes, simplify this sentence to the single word "LIKE".

```python
import random
import string
letters=" "+string.ascii_uppercase
print("Letters=",letters)
goal="I LIKE"    # our goal
n,tot=0,0
while(1):
  attempt=''.join(random.choice(letters) for x in
range(len(goal)))
  tot+=1
  if (attempt == goal):
      n +=1;
      print(n,tot,"P=",float(n)/tot)
      if (n>9): break
print("Probability=",float(n)/tot, "success=",n);
```

20.3 Appendix

In this program for chapter 3.4 About probabilities and information we calculate the probability of forming a specific complex sentence by either randomly adding a letter or substituting this letter into an existing word ("mutation"). The initial sentence is "I LIKE". Our goal is to obtain the sentence "I LIKE SNOW". This program is CPU intensive, and you should wait for a few hours to finish the calculation.

```python
import random
import string
letters=" "+string.ascii_uppercase
print("Letters=",letters)
initial="I LIKE    "  # initial word with 3 empty slots
goal="I LIKE SNOW"    # our goal
n,tot=0,0
short=len(initial)
while(1):
    L1=random.choice(letters)
    L2=random.choice(letters)
    L3=random.choice(letters)
    L4=random.choice(letters)
    tot +=1
    num = random.sample(range(0,short),4)
    attempt = initial[:num[0]] + L1 + initial[num[0] + 1:]
    attempt = attempt[:num[1]] + L2 + attempt[num[1] + 1:]
    attempt = attempt[:num[2]] + L3 + attempt[num[2] + 1:]
    attempt = attempt[:num[3]] + L4 + attempt[num[3] + 1:]
    #if (attempt.find("I LOVE")>-1): print("("+attempt+")", L1,num)
```

```
  if (attempt == goal):
      n +=1;
      print(n,tot,"P=",float(n)/tot,attempt)
      if (n>9): break
print("Probability=",float(n)/tot, "success=",n);
```

20.4 Appendix

In this code for chapter 4.3 The birthday problem we calculate the number of birthday matches among 23 people. Birthdays can be repeated. We also calculate the probability of more than one match, more than two matches, and more than three matches. If you're interested in matches between all birthdays and death days, then set SignificantDates=2 (2 significant dates).

```python
import random
import itertools as it

# how many trials to estimate probability
MaxCount=100000
# How many known people to watch
MaxPeople=23
# How many significant events
SignificantDates=1

event,event2,event3,event4=0,0,0,0
for i in range(MaxCount):
    if (i%100 ==0): print("Process=",i)
    xlist=[]
    people=[]
    for j in range( MaxPeople ):
        numb = random.sample(range(1, 366), SignificantDates)
        xlist.append(numb)
        people.append(j)

    matches=[]
    for j1, j2 in it.combinations(people, 2):
        ml1= xlist[j1]
```

```
        ml2= xlist[j2]
        for m1 in range(SignificantDates):
            for m2 in range(SignificantDates):
                if (ml1[m1] == ml2[m2]):
                    matches.append([j1,j2,m1,m2,ml1[m1],ml2[m2]])


    if (len(matches)>0): # just any 2 matches
        event=event+1
        for x in range(len(matches)):
            ev=matches[x]
            print("Match Nr =",x)
            print("Event=",event," Total=", i,"
Idx=",ev[0],ev[1]," Match =",xlist[ev[2]],xlist[ev[3]])
    if (len(matches)>1):
        event2=event2+1
        print(" ## >1 match! Event=",event2," Total=",
i,"Inx=",matches)
        for x in range(len(matches)):
            ev=matches[x]
            print(" Match Nr =",x)
            print(" Index=",ev[0],ev[1]," Match
value=",ev[4],ev[5])
    if (len(matches)>2):
                event3=event3+1
    if (len(matches)>3):
                event4=event4+1

print("Probability of >0 pair  =", float(event)/MaxCount)
print("Probability of >1 pair  =", float(event2)/MaxCount)
print("Probability of >2 pair  =", float(event3)/MaxCount)
print("Probability of >3 pair  =", float(event4)/MaxCount)
```

20.5 Appendix

For three people with two significant dates each (birth and death), we'll find a match of one person's significant date with another's. And then, a match with one of the dates of the third person. This code is used in chapter 6.2 Stephen Hawking.

```python
import random
import itertools as it
MaxCount=1000 # how many successful events to estimate proba-
bility
MaxPeople=3  # How many known people to watch
# How many significant events (death, birth)
SignificantDates=2
event,tot=0,0
while(1):
    xlist=[]
    people=[]
    for j in range( MaxPeople ):
        numb = random.sample(range(1, 366), SignificantDates)
        xlist.append(numb)
        people.append(j)
    match=0
    M1,M2,M3,J1,J2,J3=-1,-1,-1,-1,-1,-1
    for j1, j2, j3 in it.combinations(people, 3):
        ml1,ml2,ml3=xlist[j1],xlist[j2],xlist[j3]
        for m1 in range(SignificantDates):
            for m2 in range(SignificantDates):
                if (ml1[m1] == ml2[m2] ):
                    for m3 in range(SignificantDates):
                        if (ml1[m1] == ml3[m3]):
                            match=match+1

J1,J2,J3,M1,M2,M3=j1,j2,j3,m1,m2,m3
```

```
        tot +=1
        if (match>0):
            event=event+1
            if (event>MaxCount): break
            ml1,ml2,ml3=xlist[J1],xlist[J2],xlist[J3]
            print("This happened! Event=",event," In-
dex=",J1,J2,J3)
            print("Signatures=",M1,M2,M3," Match
value=",ml1[M1],ml2[M2],ml3[M3])
            print(len(xlist),xlist)
prob=float(event)/tot
print("Probability of at least 1 match =",prob," total=",tot,"
success=",event);
```

20.6 Appendix

This program code calculates the probability of 4 significant events of coincidences for chapter <u>6.5 Hitler and Napoleon</u>. To increase the precision of the calculation, increase the "MaxCount" variable. The current program gives an approximate value given in the above-mentioned chapter.

```python
import random
# how many successful events to estimate probability
MaxCount=100

# How many significant event matches
NrEvents=4

# difference between the 4 events
diff=129

# active period of life: 16 - 56 (moment they lost)
active_years=40

tot,event=0,0
while(1):
  napoleon = random.sample(range(0, active_years), NrEvents)
  napoleon.sort()
  gitler = random.sample(range(0+diff, active_years+diff),
NrEvents)
  gitler.sort()
  tot +=1
  if (gitler[0]-napoleon[0] == diff):
     if (gitler[1]-napoleon[1] == diff):
        if (gitler[2]-napoleon[2] == diff):
           if (gitler[3]-napoleon[3] == diff):
              event=event+1
```

```
            if (event>MaxCount): break
print("Probability of ",NrEvents," events=", float(event)/tot)
```

20.7 Appendix

In this application, we demonstrate how to derive the number "666" from the word "Kaiser". See chapter 6.6 The Kaiser and the War. Then, we calculate the probability that the number "666" can be associated with a random word consisting of 4 to 12 letters using 60 simple methods by assigning letters to their position in the alphabet and then performing simple manipulations (addition, subtraction, etc.).

The resulting probability is 0.007. Note that out of 10,000 random words that could somehow be associated with "666", no word resembling "Kaiser" or any known name was found. To see such a word (or a word resembling some name), much larger statistics would be required.

```python
import random

goal="KAISER"
letters="ABCDEFGHIJKLMNOPQRSTUVWXYZÄÖÜß"

# this is how to get 666 from KAISER
xsum=0
for i in goal:
    k=1+letters.find(i)
    xsum +=int(str(k)+"6" )
print("Sum=",xsum)

# randomly generate string between with the length 4-12 let-
ters,
# try 60 algorithms to create 666  from random strings
# how many finds
MaxFound=10000
```

```
print("Letters=",letters)
ev,tot=0,0
while(1):
  mylength = random.randint(4, 12) # word with length from 4
to 12
  attempt=''.join(random.choice(letters) for x in range(
mylength ))
  xsum = [0]*60   # try 60 simplest algorithms to get 666
  for i in attempt:
    k=1+letters.find(i)
    for n in range( 0,10): xsum[n] +=int(str(k)+str(n) )
    for n in range(20,30): xsum[n] +=k+n
    for n in range(30,40): xsum[n] +=k-n
    for n in range(40,50): xsum[n] +=int(k*n)
    for n in range(50,60): xsum[n] +=int(k/n )
  ev +=1
  # did you find 666 in this random string?
  if 666 in xsum:
              # print("Found=666",ev,tot, "Word=",attempt)
              tot +=1
              if (tot>MaxFound): break
print("Probability for 666 from random strings
=",float(tot)/ev)
```

20.8 Appendix

In this application, we calculate the probability that out of 21 days of observations, 4 days are solar, and these 4 days correspond to some other significant events. See the description in chapter 6.8 From my experience.

```python
import random

# how many successful events to estimate probability
MaxCount=100

# total days of observation
days=21

# expected sunny days
sunny=[0,3,8,20]

tot,event=0,0
while(1):
    random_sunny = random.sample(range(0, days), len(sunny))
    tot +=1
    s=0
    for i in sunny:
      for j in random_sunny:
          if (i == j ): s +=1
    if (s==len(sunny)):
        event +=1
        if (event>MaxCount): break
print("Probability of ",sunny," events=", float(event)/tot)
```

20.9 Appendix

This program for chapter 9.1 The number π calculates the position of the number "999999" in the constant "π". It calculates the probability of the number 999999 appearing at a position less than or equal to 762 in a large set of irrational numbers. Then the program finds all possible six-digit numbers that appear with a probability less than the probability of 999999 appearing in the number "π".

To print the number "π" to an accuracy of 770 digits, simply add "print(PI)" after the line of code where the variable PI is defined. To calculate the probability of the number "999999" appearing, we use the division of irrational numbers using a pseudo-random number generator.

This program requires the specialized library "mpmath", which can be downloaded from this link (The MPmath development team, 2023) for free.

```python
import sys,random
from itertools import product
from mpmath import mp

def calcProb(PI, find, maxdps):
  precision=20 # max found
  Position=0
  mp.dps = maxdps # precision
  try:
    Position=PI.index(find)
    if (Position>0):
        print("Find ",find," in PI at position=",Position)

            if (Position<7): return 1
  except ValueError: return 1.0 # Not found at this precisions
```

```
    if (Position>0): # reset position
        mp.dps = Position + 1;
        print("Redefine precision =",mp.dps)
    j,n,S=0,0,0
    delta=0.01
    while(1): # infinite loop to collect statistics
        M=mp.mpf( 1+delta*j*random.uniform(1.0, 2.0) )
        R=random.uniform(1, 10.0)
        S = mp.sqrt(M-delta) / (mp.sqrt(R));
        if (S>9.99):
            delta=delta*0.1
            continue
        ST=str(S)
        j +=1
        idx=-1
        try:
          idx=ST.index( find )
        except ValueError: pass # Not found
        if (idx>-1 and idx<=Position):
            if (n>=precision): break
            n +=1
    return float(n)/j


mp.dps = 770 # initial precision
PI=str(mp.pi)
find="999999"
p999999=calcProb(PI, find, mp.dps)
print("Probability for pattern ", find, " is ", p999999)
res={}
for x in product("0123456789",repeat=6):
    find="".join(x)
    p=calcProb(PI, find, mp.dps)
    if (p<p999999):
      print("Probability for pattern ", find, " is ", p)
      res[find] =p
i=0
for key, value in res.items():
```

```
    print(i, key, ":", value)
    i +=1
```

20.10 Appendix

In this program for chapter 9.2 The number *e* we find the probability that a block of any four numbers will repeat at least twice in a row in the number "*e*" (the base of the natural logarithm). First, we calculate the position of the repetition of the number 1828. This position is 9. Then, we generate many random (irrational) numbers and find the probability of any 4-digit number reappearing at a position equal to or less than position 9.

```python
import random
from mpmath import mp
# maximum number of matches
NumMax=100
# initial precision (can be arbitrary)
mp.dps = 1000
print("Base of the natural logarithm= ",str(mp.e))

# return last position of repeated value
def getRepeated4(NUM):
  pos=0
  idx=NUM.find(".") # position after .
  if (idx>-1): NUM=NUM[idx+1:]
  for i in range(len(NUM)-8):
      h1,h2=[],[]
      for j in range(i,i+4):
          h1.append(NUM[j])
          h2.append(NUM[j+4])
      if (h1[0] == h2[0] and h1[1] == h2[1] and
          h1[2] == h2[2] and h1[3] == h2[3]):
          pos=i+8
          break
  return pos
```

```
pos=getRepeated4( str(mp.e) )
print("Last position of repeated values",pos)

mp.dps=pos+2 # redefine precision
j,event=0,0
delta=0.01
while(1): # infinite loop to collect statistics
    M=mp.mpf( 1+delta*j*random.uniform(1.0, 2.0) )
    R=random.uniform(1, 10.0)
    S = mp.sqrt(M-delta) /  (mp.sqrt(R));
    ST=str(S%1) # number after decimal
    j +=1
    if (getRepeated4(ST)>0):
        event +=1
        if (event>NumMax): break
print("Prob for 1st repeated=",float(event)/j);
```

20.11 Appendix

Here we present the program for chapter 9.7 The fine structure of the world.

The first part of the code estimates the probability of finding a real number between 85 and 185 ("life is possible") which is as close to some integer number as $1/\alpha$ itself. The result of the calculation gives a probability of about 0.069.

```python
import random

nTotal=1000000
nEvent=0
# Life is possible for 85-180
alphaRange=[85,180]

def relativeError(y):
    yint=round(y)
    diff=abs(yint-y)
    return diff/yint

alphaInv=137.035999084
relAlpha=relativeError(alphaInv)
print("Relative deviation from integer=",relAlpha)
for i in range(1, nTotal):
    y = random.uniform(alphaRange[0], alphaRange[1])
    rel=relativeError(y)
    if (rel<relAlpha): nEvent += 1
print("Probability=",nEvent/float(nTotal)," found=", nEvent)
```

The second code shows the calculation of the fine structure constant using two methods and then it compares the calculations with experimental data.

```
import math

e =1.602176565e-19 # C
e0=8.854187817e-12 # F·m-1
h= 6.62607015e-34  # J·s
c= 299792458.0
pi=math.pi

alp_inv=1.0 /(e*e / (2*e0*h*c) )
alp_inv_geom=(4*pi*pi*pi + pi*pi+ pi)

print("From e,h,e,c constants =",alp_inv)
print("From Geometry using PI =",alp_inv_geom)
print("Experimentally measured=",137.035999206)
```

20.12 Appendix

This simple program calculates the probability of obtaining a height-to-base ratio consistent with the value of 0.636 within a relative error of 0.17% when randomly selecting values for the height of the Great Pyramid of Giza in the range from 100 to 150 meters. See chapter 9.9 The Great Pyramid and numbers.

```python
import random
# Average radius of the Earth and Moon
EarthR=6371000
MoonR=1737100
Ratio=(EarthR+MoonR) / (2*EarthR)
print("Ratio=",Ratio)

# average sizes of the Pyramid
Pside=230
Phight=146.6

# expected precision
Precision=0.0017

nn,tot=0,0
while(1):
    T=random.uniform(50, 150)/Pside
    if (abs((T-Ratio)/Ratio) < Precision):
        nn=nn+1
        if (nn>100): break
    tot = tot+1
print("Probability  =", float(nn)/tot)
```

20.13 Appendix

Here you will find the computational program for chapter 10.2 The Urantia Book. We provide a code written in Python that checks the day of the week for 14 days of the year (according to the Julian calendar) taken from The Urantia Book (Urantia Foundation 2008). We found 14 places in the book where the exact date of events, including the day of the week and year, is specified. Events without specified years were not used to reduce uncertainty in interpretation. Then, we used the code to verify the correctness of the weekdays. Out of the 14 randomly selected days, only one day did not match our calculations. For the position "129:1.1 (1419.4)", our calculation indicates "Thursday", not "Sunday".

```python
def weekday( data ):
    """ Julian day of week, Sunday = 1, Saturday = 7
    http://en.wikipedia.org/wiki/Zeller%27s_congruence """
    year,m,q=data[0],data[1],data[2]  # year,month,day
    if m == 1:
        m = 13; year -= 1
    elif m == 2:
        m = 14; year -= 1
    K = year % 100
    J = year // 100
    f = (q + int(13*(m + 1)/5.0) + K + int(K/4.0))
    fg = f + int(J/4.0) - 2 * J
    fj = f + 5 - J
    if year > 1582: h = fg % 7
    else: h = fj % 7
    if h == 0: h = 7
    days = ["Sunday", "Monday", "Tuesday", "Wednesday",
"Thursday", "Friday", "Saturday"]
    return days[h-1]


ur={}
ur["191:3.3 (2041.2)"]=["April 14, a.d.2", (2,4,14), "Friday"]
ur["124:3.4 (1370.2)"]=["June 24, a.d.5", (5,6,24), "Wednes-
day"]
ur["137:8.3 (1535.9)"]=["June 18, a.d.26", (26,6,18), "Tues-
day"]
ur["141:0.1 (1587.1)"]=["January 19, a.d.27", (27,1,27), "Mon-
day"]
ur["130:0.1 (1427.1)"]=["April 26, a.d.22", (22,4,26), "Sun-
day"]
ur["126:3.2 (1389.5)"]=["April 17, a.d.9", (9,4,17), "Wednes-
day"]
ur["124:6.1 (1374.1)"]=["April 9, a.d.7", (7,4,9), "Saturday"]
ur["124:5.2 (1373.2)"]=["January 9, a.d.7", (7,1,9), "Sunday"]
ur["129:1.1 (1419.4)"]=["January 9, a.d.21", (21,1,9), "Sun-
day"]
```

```
ur["135:8.2 (1503.5)"]=["January 12, a.d.26", (26,1,12), "Sat-
urday"]
ur["146:0.1 (1637.1)"]=["January 18, a.d.28", (28,1,18), "Sun-
day"]
ur["158:0.1 (1752.1)"]=["August 12, a.d.29", (29,8,12), "Fri-
day"]
ur["161:0.1 (1783.1)"]=["September 25, a.d.29", (29,9,25),
"Sunday"]
ur["149:0.1 (1668.1)"]=["October 3, a.d.28", (3,10,28), "Sun-
day"]
i=1
for k, v in ur.items():
    print(i,k, "expect=",v[2], "calculated=",weekday(v[1]))
    i +=1;
```

20.14 Appendix

For chapter <u>10.4 Premonitions of tragedies</u> we present code that includes data on the number of passengers on the days of the accidents and on the preceding, non-accident days. The data is taken from the publication (Cox 1956). The value -1 corresponds to days when information could not be obtained. We use the DataMelt program (Chekanov 2016) to construct this graph. It shows the distribution of deviations for the number of passengers from the average value, calculated for the days when there were no accidents. Indeed, the distribution is slightly shifted to the left, as would be expected if passengers could sense the tragedy. However, such a deviation is too small to confidently assert the effect of foresight.

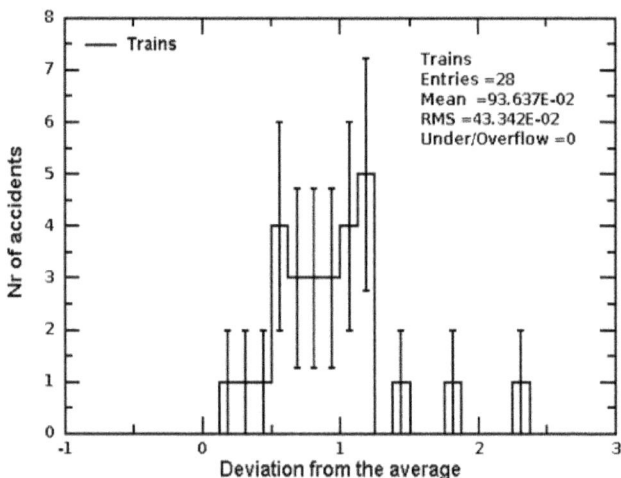

```
from jhplot import HPlot,H1D

cvs_data='''Line|Train|Date|D-7|D-6|D-5|D-4|D-3|D-2|D-1|D-ac-
cident
Boston & Maine|#60|07/05/53|36|83|53|-1|34|34|32|23
Boston & Maine|#1115|18/12/51|182|163|173|166|-1|-1|151|203
Canadian Pacific|#2960|21/02/53|123|85|62|54|46|46|80|126
Central Vermont|#332|15/07/51|173|131|94|75|124|140|187|147
Chicago & Est III.|Georgian|15/06/52|35|70|62|48|53|60|68|9
Chi. Mil., St. P. &
Pac|#15|15/12/52|150|136|100|120|87|118|86|55
Louisville & Nashville|#99|02/02/52|77|110|38|48|51|52|133|85
New York Central|#5|27/03/53|109|84|161|49|84|86|81|55
New York Central|#12|27/03/53|86|84|185|66|65|60|82|109
New York Central|#27|04/10/50|57|93|87|79|90|62|59|53
Western Maryland|#2|14/02/51|6|7|2|10|-1|7|6|9
Atlantic Coast Line|#2|20/04/53|53|50|55|52|54|41|45|52
Atchison Topeka & Santa Fe|#2|22/09/54|73|78|50|71|77|75|66|48
Atchison Topeka & Santa Fe|#4|06/09/54|11|11|14|6|12|6|8|3
Atchison Topeka & Santa
Fe|#15|13/11/53|55|61|50|38|49|70|49|63
Atchison Topeka & Santa
Fe|#19|22/08/54|43|55|44|51|58|42|52|35
Atchison Topeka & Santa
Fe|#20|24/11/54|32|41|82|46|36|60|37|40
Chicago & Est III.|Georgian|15/06/52|106|56|49|45|55|83|89|159
Chi. Mil., St. P. & Pac|#15|15/12/52|69|41|35|53|63|50|29|46
Chi. Mil., St. P. & Pac|#16|31/05/53|18|25|15|13|16|24|13|19
Chi. Mil., St. P. & Pac|#16|05/09/54|28|36|53|55|32|62|56|42
Kansas City Southern|#1|10/08/51|30|30|13|15|12|20|16|12
Louisville & Nashville|#99|10/08/51|66|37|36|33|49|62|56|25
New York  |#5|i/n/Si|45|23|43|35|26|31|44|31
New York Central|#12|3/27/53|103|40|120|100|54|69|73|96
New York Central|#27|04/10/50|129|159|143|135|109|95|77|96
Pennsylvania|#17|06/09/53|16|9|16|13|12|20|15|11
Pennsylvania|#173|15/06/53|182|132|114|56|196|176|180|161'''
```

```
c1 = HPlot("Canvas")
c1.visible(1)
c1.setAutoRange()
c1.setMarginLeft(70)
c1.setNameX("Deviation from the average")
c1.setNameY("Nr of accidents");
h1= H1D("Trains", 20, 0.0, 2.5)

import csv
reader = csv.reader(cvs_data.splitlines(), delimiter='|')
n=0;
plus=0
minus=0
for row in reader:
    n +=1
    if n<2: continue
    m,k=0,0
    nr_in_accident=float(row[len(row)-1])
    for i in range(3,len(row)-1):
            nr=float(row[i])
            if (nr>0):
                    m=m+nr
                    k +=1
    mean=m/k
    print(mean, nr_in_accident, mean/nr_in_accident)
    h1.fill(  nr_in_accident/mean )
    if (mean>nr_in_accident):  plus +=1
    if (mean<=nr_in_accident): minus +=1

print("Plus=",plus)
print("Minus=",minus)
c1.drawStatBox(h1)
stat= h1.getStat()
print stat["mean"],"+-",stat["mean_error"]
c1.draw(h1)
```

21 Bibliography

3600+ French Last Names and Meanings/ Find your French last name and learn about its meaning and origins. (2024). Retrieved January 15, 2024 from FamilyEducation: https://www.familyeducation.com/baby-names/surname/origin/french

Aczel, A. (2015). *Why Science Does Not Disprove God.* HarperCollins.

Agarwal, R., & Agarwal, H. (2021). Origin of Irrational Numbers and Their Approximations. *Computation, 9*(3), 29. From https://www.mdpi.com/2079-3197/9/3/29

Arndt, J., & Haenel, C. (2001). *Pi - Unleashed.* Springer Berlin Heidelberg.

Atwater, P. H. (2007). *The Big Book of Near-death Experiences: The Ultimate Guide to What Happens When We Die.* Hampton Roads Pub.

Avella, A. (2022, January 24). *Physics - Quantum Mechanics Must Be Complex.* Retrieved January 2, 2024 from Physics (APS): https://physics.aps.org/articles/v15/7

Balasubramanian, V. (2006, July 26). *Penn Researchers Calculate How Much The Eye Tells The Brain.* Retrieved December 17, 2023 from ScienceDaily: http://www.sciencedaily.com/releases/2006/07/060726180933.htm

Ball, W. R., & Coxeter, H. M. (1987). *Mathematical Recreations and Essays.* (H. M. Coxeter, Ed.) Dover Publications.

Barnett, T. (2015, September 16). *Building a Protein by Chance.* Retrieved February 2, 2024 from Stand to Reason: https://www.str.org/w/building-a-protein-by-chance

Barrow, J. D. (2001). Cosmology, life, and the anthropic principle. *nnals of the New York Academy of Sciences, 950 (1)*, 139–153.

Beitman, B. (2022). *Meaningful Coincidences: How and Why Synchronicity and Serendipity Happen.* Inner Traditions/Bear.

Beitman, B. B., Celebi, B. D., & Elif Coleman, S. L. (n.d.). Synchronicity and healing. *Integrative psychiatry (Oxford University Press.)*, 445–483.

Belousov, L. S., & Manykin, A. (Eds.). (2014). *Первая мировая война и судьбы европейской цивилизации.* Izdat. Moskovskogo Univ. Retrieved February 11, 2024

Ben, D., Tressoldi, P. E., Rabeyron, T., & Duggan, M. (2016). Feeling the future: A meta-analysis of 90 experiments on the anomalous anticipation of random future events. *F1000Research.* From https://f1000research.com/articles/4-1188/v2

Benedictus, L. (2020). *The Great Pyramid's location isn't as spooky as this post makes out.* From Full Fact.: https://fullfact.org/online/great-pyramid-speed-of-light/

Berengut, J. C., Flambaum, V. V., King, J. A., & Webb, J. K. (2011). Is there further evidence for spatial variation of fundamental constants? *Phys. Rev. D., 83*, 123506.

Borbley, A. (1988). *Secrets Of Sleep.* (D. Schneider, Trans.) Basic Books. Retrieved March 9, 2024

Boyer, A. (2019). Weathering the Storm: Physiological and Behavioural Responses of White-Throated Sparrows to Inclement Weather Cues. *Electronic Thesis and Dissertation Repository*, 6215.

Browne, S., & Harrison, L. (2008). *End of Days: Predictions and Prophecies about the End of the World.* Dutton.

Cambray, J. (2012). *Synchronicity: Nature and Psyche in an Interconnected Universe.* Texas A&M University Press. Retrieved December 14, 2023

Chekanov, S. V. (2016). *Numeric Computation and Statistical Data Analysis on the Java Platform.* Springer International Publishing. Retrieved December 17, 2023

Chopra, D. (2005). *Synchrodestiny.* Rider.

Colburn, T. A. (2000). Information, thought, and knowledge. *In Proceedings of the world multiconference on systemics, cybernetics and informatics*, 467—471.

Cox, W. E. (1956). Precognition: An analysis. II. Subliminal precognition. *Journal of the American Society for Psychical Research, 55*, 99-109.

Csanyi, E. (2012, September 12). *Nikola Tesla - Everything is the Light.* Retrieved February 16, 2024 from Electrical Engineering Portal: https://electrical-engineering-portal.com/nikola-tesla-everything-is-the-light

Dare, L. (2017). *Biblical Numerology: Meaningful Numerical Values and Patterns of the Holy Bible.* CreateSpace Independent Publishing Platform. Retrieved February 16, 2024

Darwin, C. (1979). *The origin of species : complete and fully illustrated. Original publication year 1859.* Gramercy Books.

Davies, O. (2018). *A Supernatural War: Magic, Divination, and Faith During the First World War.* (O. Davies, Ed.) Oxford University Press. Retrieved February 12, 2024

Dawkins, R. (2006). *The Selfish Gene.* OUP Oxford. Retrieved January 31, 2024

Dawkins, R. (2008). *The God Delusion.* Houghton Mifflin Company.

Dimitrov, T. (1995). *50 Nobel Laureates and Other Great Scientists who Believe in Go.* Novelists.net.

Dobrogosz, H. (2024, January 31). *40 Coincidences That Prove The World Is Small.* Retrieved February 2, 2024 from BuzzFeed: https://www.buzzfeed.com/hannahdobro/shocking-small-world-stories

Ėkshtut, S. A. (1994). *В поиске исторической альтернативы: Александр I, его сподвижники, декабристы.* Россия молодая.

Faggin, F. (2024). *Irreducible: Consciousness, Life, Computers, and Human Nature.* Essentia Books.

Feuillet, L., Dufour, H., & Pelletier, J. (2007). Brain of a white-collar workeк. *The Lancet, 390,* 262.

Feynman, R. P., & Leighton, R. (1997). *"Surely you're joking, Mr. Feynman!" : adventures of a curious character.* (R. Leighton, & E. Hutchings, Eds.) W.W. Norton.

Geisler, N. L., & Turek, F. (2004). *I Don't Have Enough Faith to Be an Atheist.* Crossway.

Geison, G. L. (2014). *The Private Science of Louis Pasteur.* Princeton University Press.

Ginsburgh, I., & Taylor, G. L. (1987). *Scientific Predictions of The Urantia Book.* Retrieved February 8, 2024 from The Urantia Book and Contemporary Christian Beliefs About Revelation: https://archive.urantiabook.org/archive/science/ginsss2.htm

Gooding, D. W., & Lennox, J. C. (2018). *Suffering Life's Pain: Facing the Problems of Moral and Natural Evil.* Myrtlefield House.

Gorvett, Z., & Brunelle, F. (2016, July 13). *You are surprisingly likely to have a living doppelganger.* Retrieved February 14, 2024 from BBC: https://www.bbc.com/future/article/20160712-you-are-surprisingly-likely-to-have-a-living-doppelganger

Hancock, G. (1995). *Fingerprints of the Gods: The Evidence of Earth's Lost Civilization.* Crown.

Hands, J. (2016). *Cosmosapiens: Human Evolution from the Origin of the Universe.* Harry N. Abrams. Retrieved January 31, 2024

Hawking, S. W. (1988). *A brief history of time : from the big bang to black holes.* Bantam.

Hawking, S., & Mlodinow, L. (2010). *The Grand Design.* Random House Publishing Group.

Hermanns, W. (1983). *Einstein and the poet : in search of the cosmic man.* Branden Press.

Hofstadter, D. (1985). *Metamagical Themas.* Basic Books.

Howells, K. (2023, February 7). *Our favorite moons of the Solar System.* Retrieved March 17, 2024 from The Planetary Society: https://www.planetary.org/articles/our-favorite-moons-of-the-solar-system

Husain, A., Reddy, J., & Sajid, M. (2021). Fractal dimension of coastline of Australia. *Scientific Reports, 11.*

Impey, C. (2012). *How It Began: A Time-Traveler's Guide to the Universe.* W. W. Norton.

Jung, C. G. (1901). *The Symbolic Life: Miscellaneous Writings (The Collected Works of C. G. Jung, Volume 18) (The Collected Works of C. G. Jung, 68) Hardcover.* Princeton University Press.

Jung, C. G. (1973). *Synchronicity: An Acausal Connecting Principle.* Princeton University Press. Retrieved December 14, 2023

Jung, C. G., & Pauli, W. E. (2012). *The Interpretation of Nature and the Psyche.* Ishi Press International. Retrieved December 14, 2023

Kanakogi, Y., & others. (2022). Third-party punishment by preverbal infants. *Nat Hum Behav, 6,* 1234–1242.

Kittel, W., & De Wolf, E. A. (2005). *Soft Multihadron Dynamics.* World Scientific. Retrieved February 4, 2024

Kline, M. (1972). *Mathematical Thought from Ancient to Modern Times.* Oxford University Press; Illustrated edition.

Krauss, L. M. (2013). *A Universe from Nothing: Why There Is Something Rather than Nothing.* Atria.

Lange, M. (2010). What Are Mathematical Coincidences (and Why Does It Matter)? *Mind, 119*, 474.

Lennox, J. C. (2009). *God's Undertaker: Has Science Buried God?* Lion. Retrieved January 28, 2024

Lennox, J. C. (2019). *Can Science Explain Everything?* Good Book Company. Retrieved January 18, 2024

Lennox, J. C. (2021). *God and Stephen Hawking 2ND EDITION: Whose Design Is It Anyway?* Lion Hudson PLC.

Lim, M. (2021, July 30). *Turning DNA into data storage powerhouses.* Retrieved December 22, 2023 from DUG Technology: https://dug.com/turning-dna-into-data-storage-powerhouses/

Mandelbrot, B. B. (1983). *The fractal geometry of nature.* Henry Holt and Company. Retrieved February 4, 2024

Masci, D. (2009, November 5). Religion and Science in the United States. *Pew Research Center.* Retrieved December 20, 2023 from https://www.pewresearch.org/religion/2009/11/05/an-overview-of-religion-and-science-in-the-united-states/

Menskiĭ, M. B. (2011). *Сознание и квантовая механика: жизнь в параллельных мирах : (чудеса сознания-из квантовой реальности) : [пер. с англ.].* Век 2.

Meyer, S. C. (2009). *Signature in the Cell: DNA and the Evidence for Intelligent Design.* HarperCollins.

Meyer, S. C. (2013). *Darwin's Doubt: The Explosive Origin of Animal Life and the Case for Intelligent Design.* HarperCollins.

Meyer, S. C. (2021). *The Return of the God Hypothesis: Three Scientific Discoveries Revealing the Mind Behind the Universe.* HarperOne.

Miller, B. (2019, February 18). *A Dentist in the Sahara: Doug Axe on the Rarity of Proteins Is Decisively Confirmed.* Retrieved February 2, 2024 from Evolution News: https://evolutionnews.org/2019/02/a-dentist-in-the-sahara-doug-axe-on-the-rarity-of-proteins-is-decisively-confirmed/

Miller, J. (2022). Does quantum mechanics need imaginary numbers? *Physics Today, 75*(3), 14-16.

Mindell, A. (2000). *Quantum Mind: The Edge Between Physics and Psychology.* Lao Tse Press. Retrieved February 2, 2024

Moody, R. A. (2005). *The Light Beyond.* Rider.

Murray, D. B., & Teare, S. W. (1993, 11). Probability of a tossed coin landing on edge. *NASA/ADS.* Retrieved December 21, 2023 from https://ui.adsabs.harvard.edu/abs/1993PhRvE..48.2547M/abstract

Mushtaq, H. (2023). *You'll be amazed by how much data DNA can store (figures inside)! — Steemit.* Retrieved December 22, 2023 from Steemit: https://steemit.com/science/@hmushtaq/you-ll-be-amazed-by-how-much-data-dna-can-store-figures-inside

Nelson, G. K. (1969). *Spiritualism and Society.* Routledge & K. Paul.

Nucleic acid memory. (2016, November 7). Retrieved December 22, 2023 from Tature: https://www.nature.com/articles/nmat4594

Penrose, R. (1989). *The emperor's new mind concerning computers, minds, and the laws of physics.* Oxford University Press.

Penrose, R. (2007). *The Road to Reality: A Complete Guide to the Laws of the Universe.* Knopf Doubleday Publishing Group.

Peoc'h, R. (1995). Psychokinetic Action of Young Chicks on the Path of an Illuminated Source. *Journal of Scientific Exploration, 9,* 223.

Pietsch, P. (1981). *Shufflebrain.* Houghton Mifflin.

Ponte, D. M., & Schäfer, L. (2013, Nov). Carl Gustav Jung, Quantum Physics and the Spiritual Mind: A Mystical Vision of the Twenty-First Century. *Behav Sci (Basel)., 3*(4), 601-618.

Roskies, R., & Peres, A. (1971). A new pastime–calculating alpha to one part in a million. *Phys. Today, 24,* 9.

Sanger, L. (2024). *God exists. A Philosophical Case for the Christian God.* Sanger Press.

Schäfer, L. (2013). *Infinite Potential: What Quantum Physics Reveals about how We Should Live.* Deepak Chopra Books.

Sellers, J. (2017). Out-of-Body Experience: Review & a Case Study. *Psychology, Medicine Journal of Consciousness Exploration & Research.*

Shalev, B. A. (2002). *100 Years of Nobel Prizes.* Americas Group. Retrieved January 21, 2024

Shermer, M. (2018). *Heavens on Earth: The Scientific Search for the Afterlife, Immortality, and Utopia.* Henry Holt and Company.

Social Security. United States government. (2020). *Actuarial Life Table. Statistical tables, www.ssa.gov.* Retrieved 2023 from www.ssa.gov: https://www.ssa.gov/oact/STATS/table4c6.html

Swinburne, R. (2004). *The Existence of God.* Clarendon Press.

Taleb, N. N. (2008). *The Black Swan: The Impact of the Highly Improbable.* Random House Publishing Group.

Taylor, G. (2017, November 9). Scientific Predictions of the Urantia Book. Retrieved February 8, 2024 from http://evolving-souls.org/wp-content/uploads/2018/02/Scientific-Assertions-Predictions-of-UB.pd

The MPmath development team. (2023). *MPmath.* Retrieved December 30, 2023 from mpmath - Python library for arbitrary-precision floating-point arithmetic: http://mpmath.org/

Urantia Foundation. (2008). *The Urantia Book.* (Urantia Foundation, Ed.) Urantia Foundation.

Von Daniken, E. (1999). *Chariots of the Gods.* Penguin Publishing Group.

Weber, R. (1990). *Dialogues with Scientists and Sages: The Search for Unity.* Penguin Group (USA) Incorporated.

Wells, D. (1988). Which is the most beautiful?. *The Mathematical Intelligencer, 10*, 30-31.

Wheeler, J. (1990). *Complexity, Entropy And The Physics Of Information.* (W. H. Zurek, Ed.) Avalon Publishing. Retrieved February 13, 2024

Wolfram, S. (2002). *A new kind of science.* Wolfram Media.

Yee, J. (2019, May 20). *The Relationship of the Fine Structure Constant and Pi, viXra.org e-Print archive.* Retrieved February 15, 2024 from viXra.org: https://vixra.org/abs/1905.0396

Балашов, Л. (2022). *Случайность как категория философии | Лев Балашов. Философия, политика.* Retrieved February 9, 2024 from Дзен: https://dzen.ru/a/Yz3eyFXW9m6go7Mh

Блаватская, Е. П. (1888). *Тайная доктрина («The Secret Doctrine»).* Эксмо (2008) Переводчик: Рерих Е. Редактор: Галий В.

Карышева, И. А. (1895). *Основы истинной науки - Книга 1-я Бог не опровержим наукой.* Книга.

Налимов, В. В., & Дрогалина, Ж. А. (1995). *Реальность нереального: вероятностная модель бессознательного.* Мир Идей. АО AKRON.

Пирогов, Н. И. (2010). *Вопросы жизни. Дневник старого врача. Юбилейное издание к 200-летию со дня рождения Н.И. Пирогова.* Глаголъ.

Щербаков, А. (2021). *Феномен сознания вне тела.* Ridero.

22 Index

1

4

6

8

9

A

B

C

D

E

F

I

J

K

L

M

N

O

P

Q

R

S

X

Z

www.ingramcontent.com/pod-product-compliance
Lightning Source LLC
Chambersburg PA
CBHW060849120626
46553CB00001B/28